PALEOMAGNETIC PRINCIPLES AND PRACTICE

MODERN APPROACHES IN GEOPHYSICS

VOLUME 18

Managing Editor

PALEOMAGNETIC PRINCIPLES AND PRACTICE

by

LISA TAUXE

Scripps Institution of Oceanography,
La Jolla, CA, U.S.A.

KLUWER ACADEMIC PUBLISHERS

DORDRECHT / BOSTON / LONDON

Library of Congress Cataloging-in-Publication Data

Additional material to this book can be downloaded from http://extras.springer.com.

ISBN 1-4020-0850-3 (PB 2002)
ISBN 0-7923-5258-0 (HB 1998, 2002)

Published by Kluwer Academic Publishers,
P.O. Box 17, 3300 AA Dordrecht, The Netherlands.

Sold and distributed in North, Central and South America
by Kluwer Academic Publishers,
101 Philip Drive, Norwell, MA 02061, U.S.A.

In all other countries, sold and distributed
by Kluwer Academic Publishers,
P.O. Box 322, 3300 AH Dordrecht, The Netherlands.

CD-ROM image: courtesy of Gary A. Glatzmaier,
Los Alamos National Laboratory

Printed on acid-free paper

TABLE OF CONTENTS

Table of Contents

Table of Contents

Table of Contents

CD-ROM with Paleomagnetic Data Reduction software pmag 1.7:
inside back cover

PREFACE

1 The purpose of the book

Paleomagnetic data are useful in many applications in Earth Science from determining paleocurrent directions to analyzing the long-term behavior of the geomagnetic field. Despite the diversity of applications, the techniques required to obtain and analyze the data are similar. This book attempts to draw together the various principles and practices within paleomagnetism in a consistent and up-to-date manner. It was written for several categories of readers:

1) for Earth Scientists who use paleomagnetic data in their research,
2) for students taking a class with paleomagnetic content, and
3) for other professionals with an interest in paleomagnetic data.

There are a number of excellent references on paleomagnetism and on the related specialties (rock magnetism and geomagnetism). For an excellent introductory text on paleomagnetism, the reader is encouraged to explore Butler [1992]. For in-depth coverage of rock magnetism, I recommend Dunlop and Özdemir [1997]. Similarly for geomagnetism, please see Backus et al. [1996]. A rigorous analysis of the statistics of spherical data is given by Fisher et al. [1987]. The details of paleomagnetic poles are covered in van der Voo [1993] and magnetostratigraphy is covered by Opdyke and Channell [1996]. My book is intended to augment or distill information from the broad field of paleomagnetism, complementing the existing body of literature.

This book requires a knowledge of basic Earth Science, as well as some physics and statistical theory. Also, access to a UNIX or LINUX computing platform is desirable. One need not understand every detail of the book to be able to make use of the techniques in many applications, however.

2 What is in the book

Chapter 1 begins with a review of the physics of magnetic fields. Maxwell's equations are introduced where appropriate and the magnetic units are derived from first principles. The conversion of units between cgs and SI conventions is also discussed and summarized in a handy table. The book then turns to the Earth's magnetic field, discussing the geomagnetic potential, geomagnetic elements, and the geomagnetic reference fields. The various magnetic poles of the Earth are also introduced. Finally, Chapter 1 briefly describes the ancient magnetic field.

Chapter 2 deals with rock magnetism. The most important aspect of rock magnetism to the working paleomagnetist is how rocks can become magnetized and how they can stay that way. In order to understand this, the chapter begins with a discussion of what the origin of magnetism is in rocks, including induced and remanent magnetism, anisotropy energy, and mechanisms of magnetization. The magnetic properties of geologically important minerals are described and tabulated, as well as tools for their recognition.

The nuts and bolts of how to obtain paleomagnetic samples and how to treat them in the laboratory is the topic of Chapter 3. It covers sampling strategy and routine laboratory procedures, including measurement and demagnetization. Techniques for obtaining both directional and paleointensity information are outlined.

Chapter 4 describes what to do with directional data. It begins with a thorough discussion of Fisher [1953] statistics, describing many useful tests. The chapter then illustrates how to determine if a particular data set is likely to be Fisherian and introduces a bootstrap procedure for treating data that are not. Included are many bootstrap tests that perform similar functions to the Fisher based tests that have proved so useful over the years.

Magnetic tensor data, primarily anisotropy of magnetic susceptibility are useful in a number of geological applications. The acquisition and treatment of such data is described in Chapter 5. Traditional (Hext [1963]) and more modern (bootstrap) approaches are presented in some detail. Tests for discrimination of eigenparameters are developed and illustrated with many examples.

The final chapter, Chapter 6, provides a whistle-stop tour of various paleomagnetic applications. It includes discussion of magnetostratigraphy, paleointensity studies, apparent polar wander. Also included are examples of how to use magnetic tensor data for investigations on sedimentary, igneous and metamorphic rocks.

Two appendices are included in the book. The first is a description of the companion public domain software package pmag1.0, used in the examples at the back of every chapter. Appendix 1 describes how to get the programs and how to use them. Because the programs are designed to take advantage of many features of the UNIX operating system, there is also a section on "survival UNIX". Appendix 2 is a tabulation of many terms, acronyms and constants used throughout the book. It is there that the reader will find the meaning of many of the acronyms of which paleomagnetists are so fond.

3 How to use the book

The reader is encouraged to begin with Appendix 1. The utility of the book will be greatly enhanced by successfully installing and using the programs and examples at the end of each chapter. These are numbered and are referred to in the body of each chapter. By working through the examples, the reader will gain familiarity not only with UNIX and the pmag1.0 software package, but also with the concepts discussed in the chapters. Each chapter builds on the principles outlined in the previous chapters, so the reader is encouraged to work through the book sequentially.

I have attempted to maintain a consistent notation throughout the book. Vectors, axes and tensors are in bold face; other parameters, including vector components, are in italics. The most important paleomagnetic variables are listed in Appendix 2.

4 Acknowledgements

Although I am the sole author, I am indebted to many people for assistance great and small. Thanks are offered to Cathy Constable, Jeff Gee, Bob Parker and the late Geoff Watson for many of the ideas in this book. Thom Pick, Leonardo Masoni and Jeff Gee contributed unpublished data sets which greatly enhanced the book. I am grateful for the many pairs of eyes that hunted down the errors (and hopefully caught them all) in the text and the programs: Cathy Constable, Brad Clement, Jeff Gee, Ton van Hoof, Bernie Housen, Jan Klingen, Yvo Kok, Luca Lanci, Cor Langereis, David Lowe, Chad McCabe, Tom Mullender, Andrew Newell, Bob Parker, Peter Selkin, and Andy Roberts. Acknowledgement is owed to the many sources of public domain software that ended up in the package pmag1.0, including: Phil McFadden, Jeff Gee, Cathy Constable, Steve Constable, Charlie Barton. Special thanks go to Bob Parker and Loren Shure who gave the world the gift of **plotxy**. When I first used plotxy (before commercial plotting software could be had), I sang to myself the words to Amazing Grace, "I was blind, but now I see". I am also grateful for the authors of the books which both educated and inspired me which are too many to enumerate, but which are listed in the bibliography. Several people deserve special mention for assistance of a more personal kind: Carmen Luna and Hubert Staudigel. Thanks go to the people at Fort Hoofddijk who offered refuge for the long leave of absence from Scripps Institution of Oceanography that spawned this book. Finally, I would like to acknowledge two pearls of wisdom gleaned from acquaintances along the way. From Lou Jacobs, a paleontologist: "Don't get it right, get it written" and from George Clark, a carpenter: "It's what it is!"

Lisa Tauxe
Scripps Institution of Oceanography
La Jolla, CA 92093-0220
U.S.A.

Chapter 1

GEOMAGNETISM

Paleomagnetism is the study of the magnetic properties of rocks. It is one of the most broadly applicable disciplines in geophysics, having uses in diverse fields such as geomagnetism, tectonics, paleoceanography, volcanology, paleontology, and sedimentology. Although the potential applications are varied, the fundamental techniques are remarkably uniform. Thus, a grounding in the basic tools of paleomagnetic data analysis can open doors to many of these applications. One of the underpinnings of paleomagnetic endeavors is the relationship between the magnetic properties of rocks and the Earth's magnetic field. In this chapter, we will briefly introduce aspects of geomagnetism that are fundamental to paleomagnetism, including the present field and its behavior through geological time. There are many useful textbooks on geomagnetism and/or paleomagnetism that are of general interest such as McElhinny [1973], Butler [1992], van der Voo [1993], Blakely, [1995], Opdyke and Channell [1996], Merrill et al. [1996], and Backus et al. [1996]. Here we briefly sketch an outline of geomagnetism.

We will start with a review of the physics of magnetic fields in general. For excellent background reading, see Jiles [1991] and Aharoni [1996].

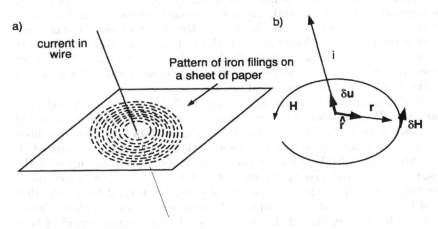

Figure 1.1. a) Distribution of iron filings on a flat sheet pierced by a wire carrying a current *i*. b) Relationship of magnetic field **H** to current *i*. $\hat{r}$ is the unit vector in the direction of **r**. δ**H** is the increment in **H** produced by the incremental length of wire δ**u** carrying current.

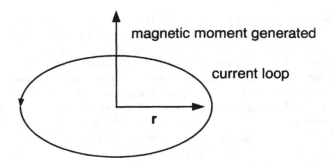

Figure 1.2. A magnetic moment is generated by current flowing in a loop.

1.1. The Physics of Magnetism

1.1.1. MAGNETIC FIELDS

Magnetic fields, like gravitational fields, cannot be seen or touched. We can feel the pull of the Earth's gravitational field on ourselves and the objects around us, but we do not experience magnetic fields in such a direct way. We know of the existence of magnetic fields by their effect on objects such as magnetized pieces of metal, naturally magnetic rocks such as lodestone, or temporary magnets such as copper coils that carry an electrical current. If we place a magnetized needle on a cork in a bucket of water, it will slowly align itself with the local magnetic field. Turning on the current in a copper wire can make a nearby compass needle jump. Observations like these led to the development of the concept of magnetic fields.

In classical electromagnetism, all magnetic fields are the result of electric currents. We can define what we mean by "magnetic fields" in terms of the electric currents that generate them. Figure 1.1a is a sketch of what might happen if we pierced a flat sheet with a wire carrying a current. If iron filings are sprinkled on the sheet, the filings would line up with the magnetic field produced by the current in the wire. A circle tangential to the field is shown to the right, which illustrates the *right-hand rule*, that is, if your right thumb points in the direction of current flow, your fingers will curl in the direction of the magnetic field. The magnetic field **H** is at right angles to both the direction of current flow and to the radial vector **r**. The magnitude of **H** is proportional to the strength of the current i. In the simple case illustrated in Figure 1.1, the magnitude of **H** is given by Ampère's law:

$$H = \frac{i}{2\pi r}.$$

The more general case (known as the Biot-Savart law) in which the wire need not be straight is given by:

$$\delta\mathbf{H} = \frac{1}{4\pi r^2}i\delta\mathbf{u} \times \hat{\mathbf{r}},$$

where $\delta\mathbf{H}$ is the incremental magnetic field caused by incremental length of wire $\delta\mathbf{u}$. $\hat{\mathbf{r}}$ is the unit vector along $\mathbf{r}$. The Biot-Savart law is equivalent to Ampère's law and also to one of Maxwell's equations of electromagnetism. In a steady electrical field, $\nabla \times \mathbf{H} = \mathbf{J}_f$, where $\mathbf{J}_f$ is the electric current density. In English, we say that the curl (or circulation) of the magnetic field is equal to the current density. The origin of the term "curl" for the cross product of the gradient operator with a vector field is suggested in Figure 1.1 in which the iron filings seem to curl around the wire.

1.1.2. MAGNETIC FLUX

The flux of a vector field (be it flowing water, wind, or a magnetic field) is the integral of the vector over a given area. Magnetic fields in free space generate magnetic flux. Magnetic Flux can be quantified when a source of flux passes through a closed circuit because it produces a current which can be measured. The density of flux lines is one measure of the strength of the magnetic field called the magnetic induction $\mathbf{B}$. Magnetic induction can be thought of as something that creates an observable torque ($\mathbf{u} \times \mathbf{B}$) on a length of wire ($\delta\mathbf{u}$) carrying an electric current (i). Similarly, the torque $\mathbf{m} \times \mathbf{B}$ is what causes the compass needle with magnetic moment $\mathbf{m}$ to jump when you turn on the current in a nearby wire (and consequently produce a magnetic induction $\mathbf{B}$). A force of 1 newton per meter is generated by passing a current of 1 ampere perpendicular to the direction of a magnetic induction of one tesla. Thus the tesla (T) is equivalent to $1 \ \mathrm{NA^{-1}m^{-1}}$. Reducing these units to their most fundamental form we find that $1 \ \mathrm{T} = 1 \ \mathrm{V \cdot s \cdot m^{-2}}$.

The volt-second is a unit in its own right, the weber (Wb), which logically must be a unit of magnetic flux. The weber is defined as the amount of magnetic flux which, when passed through a one-turn coil of conductor carrying a current of one ampere, produces an electromotive force of one volt. This definition suggests a means to measure the strength of magnetic induction and is the basis of the "fluxgate" magnetometer.

1.1.3. MAGNETIC MOMENT

We noted that an electrical current in a wire produces a magnetic field that curls around the wire. A current loop surrounding an area πr^2 and carrying a current i, as shown in Figure 1.2, creates what is called a *magnetic moment* m whose magnitude is $i\pi r^2$ and so has units of Am^2. A magnetic moment making an angle θ with a magnetic field vector H has a *magnetostatic energy* associated with it. This energy is given by $\mathbf{m} \cdot \mu_0 \mathbf{H}$ or $m\mu_0 H \cos\theta$ where m and H are the magnitudes of m and H, respectively and μ_0 is the magnetic permeability of free space (see Table 1.1).

Magnetization M is a moment per unit volume (units of Am^{-1}) or per unit mass (Am^2kg^{-1}). Sub-atomic charges such as protons and electrons can be thought of as tracing out tiny circuits and behaving as tiny magnetic moments. They respond to external magnetic fields and give rise to a magnetization that is proportional to them. The relationship between M in the material and the external field H is defined as:

$$\mathbf{M} = \chi_b \mathbf{H}. \tag{1.1}$$

The parameter χ_b is known as the *bulk magnetic susceptibility* of the material and can be a complicated function of orientation, temperature, state of stress and applied field (see Chapter 5).

Certain materials can produce magnetic fields in the absence of external electric currents. As we shall see in Chapter 2, these so-called "spontaneous" magnetic moments are also the result of spins of electrons which, in some crystals, act in a coordinated fashion, thereby producing a net magnetic field. The resulting magnetization can be fixed by various mechanisms and can preserve records of ancient magnetic fields. This *remanent magnetization* forms the basis of the field of paleomagnetism and will be discussed at length in the rest of this book.

From the foregoing discussion, we see that B and H are closely related. In paleomagnetic practice, both B and H are referred to as the magnetic field. Strictly speaking, B is the induction and H is the field, but the distinction is often blurred. The relationship between B and H is given by:

$$\mathbf{B} = \mu_o(\mathbf{H} + \mathbf{M}). \tag{1.2}$$

In the SI system, μ_o has dimensions of henries per meter and is given by $\mu_o = 4\pi \times 10^{-7} H \cdot m^{-1}$. In cgs units, μ_o is unity and H is identical to B in free space. Because SI units have only recently become the standard in paleomagnetic applications, the loose usage of B and H may perhaps be forgiven.

Magnetic fields are different from electrical fields in that there is no equivalent to an isolated electrical charge, there are only pairs of "opposite charges", or magnetic *dipoles*. An isolated electrical charge produces electrical fields that begin at the source (the charge) and diverge outward. This property of the vector field can be quantified by the "divergence" $\nabla \cdot \mathbf{E}$. As there is no equivalent magnetic source, the magnetic field has no divergence. Thus, we have another of Maxwell's equations: $\nabla \cdot \mathbf{B} = 0$.

1.1.4. THE MAGNETIC POTENTIAL

Because the curl of the magnetic field ($\nabla \times \mathbf{H}$) is not generally zero, but depends on the current density and the time derivative of the electric field, magnetic fields cannot generally be represented as gradient of a scalar field. However, in the special case away from currents and changing electric fields, the magnetic field can be written as the gradient of a scalar field that is known as the *magnetic potential V_m*, *i.e.*,

$$\mathbf{H} = -\nabla V_m.$$

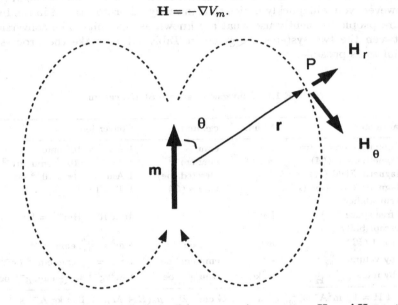

Figure 1.3. Field $\mathbf{H}$ produced at point P by a magnetic moment $\mathbf{m}$. $\mathbf{H}_r$ and $\mathbf{H}_\theta$ are the radial and tangential fields respectively.

The presence of a magnetic moment $\mathbf{m}$ creates a magnetic field which is the gradient of a scalar field (Figure 1.3). This scalar field is a function of distance from the moment. Given a *dipole moment* $\mathbf{m}$, the potential of the magnetic field produced by $\mathbf{m}$ is:

$$V_m = \frac{\mathbf{m} \cdot \mathbf{r}}{4\pi r^3} = \frac{m \cos \theta}{4\pi r^2}. \tag{1.3}$$

The radial and tangential components of $\mathbf{H}$ at P (Figure 1.3) are:

$$H_r = -\frac{\partial V_m}{\partial r} = \frac{1}{4\pi} \frac{2m \cos \theta}{r^3},$$

and

$$H_\theta = -\frac{1}{r}\frac{\partial V_m}{\partial \theta} = \frac{m\sin\theta}{4\pi r^3},$$

respectively. The units of **H** are moment per unit volume which reduce to Am^{-1}, as shown earlier.

1.2. Units conversion and confusion

We have derived magnetic units in terms of the Système International (SI). However, you will quickly notice that in many laboratories and in the literature people frequently use what are known as cgs units. The conversions between the two systems are given in Table 1.1 to make the process as painless as possible.

TABLE 1.1. Conversion between SI and cgs units.

Parameter	SI unit	cgs unit	Conversion
Magnetic moment (m)	Am^2	emu	$1\ A\ m^2 = 10^3$ emu
Magnetization (**M**)	Am^{-1}	emu cm^{-3}	$1\ Am^{-1} = 10^{-3}$ emu cm^{-3}
Magnetic Field (**H**)	Am^{-1}	Oersted (oe)	$1\ Am^{-1} = 4\pi \times 10^{-3}$ oe
Magnetic Induction (**B**)	T	Gauss (G)	$1\ T = 10^4$ G
Permeability of free space (μ_0)	Hm^{-1}	1	$4\pi \times 10^{-7}$ Hm$^{-1} = 1$
Susceptibility (χ)			
total ($\frac{\mathbf{m}}{\mathbf{H}}$)	m^3	emu oe^{-1}	$1\ m^3 = \frac{10^6}{4\pi}$ emu oe^{-1}
by volume ($\frac{\mathbf{M}}{\mathbf{H}}$)	-	emu cm^3 oe^{-1}	1 S.I. $= \frac{1}{4\pi}$ emu cm^{-3} oe^{-1}
by mass ($\frac{\mathbf{m}}{m}\cdot\frac{1}{\mathbf{H}}$)	m^{3}kg^{-1}	emu g^{-1} oe^{-1}	$1\ m^3$kg$^{-1} = \frac{10^3}{4\pi}$emu g^{-1} oe^{-1}

$1\ H = kg\ m^2A^{-2}s^{-2}$, $1\ emu = 1\ G\ cm^3$, $B = \mu_o(H+M)$, $1\ T = kg\ A^{-1}\ s^{-2}$

1.3. The Earth's Magnetic Field

One of the principal goals of paleomagnetism is to study ancient geomagnetic fields. Here, we review the general properties of the Earth's present magnetic field. The geomagnetic field is generated by convection currents in the liquid outer core of the Earth which is composed of iron, nickel and some unkown lighter component(s). Motions of the conducting fluid, which are partially controlled by the spin of the Earth about its axis, act as a self-sustaining dynamo and create an enormous magnetic field. To first order, the field is very much like one that would be produced by a gigantic bar magnet located at the Earth's center and aligned with the spin axis. In

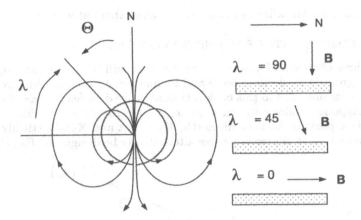

Figure 1.4. Geocentric dipole field. λ and θ are latitude and co-latitude respectively. To the right: inclinations of magnetic field **B** relative to the Earth's surface at various latitudes.

Figure 1.4, we show a cross section of the Earth with a dipolar magnetic field superimposed. If the field were actually that of a *geocentric axial dipole* (GAD), it would not matter which cross section we chose because such a field is rotationally symmetric about the axis going through the poles; in other words, the magnetic field lines would always point North. The angle between the field lines and the horizontal at the surface of the Earth, however, would vary between zero at the equator and 90° at the poles. Moreover, the magnetic field lines would be more crowded together at the poles than at the equator (the magnetic flux is higher at the poles) resulting in a polar field that would have twice the intensity of the equatorial field.

This so-called *dipole model* will serve as a useful crutch throughout our discussions of paleomagnetic data and applications, but as will be pointed out in more detail, it is a poor physical representation for what is actually producing the magnetic field.

Before looking at global maps of the geomagnetic field, we will first consider the properties of the magnetic field at a point on the surface of the Earth. The geomagnetic field is a vector, hence has both direction and intensity (see Figure 1.5). A vector in three dimensions requires three parameters to define it fully no matter what coordinate system you choose. In cartesian coordinates these would be, for example, x_1, x_2 and x_3. Depending on the particular problem at hand, some coordinate systems are more suitable to use because they have the symmetry of the problem built into them. We will be using several coordinate systems in addition to the cartesian one. We will need to convert among them at will.

1.3.1. COMPONENTS OF MAGNETIC VECTORS

The three elements of a magnetic vector that will be used most frequently are magnitude B, declination D and inclination I, as shown in Figure 1.5. The convention used in this book is that axes are denoted $\mathbf{X_1}, \mathbf{X_2}, \mathbf{X_3}$, while the components along the axes are x_1, x_2, x_3. In the geographic frame of reference, positive $\mathbf{X_1}$ is to the north, $\mathbf{X_2}$ is east and $\mathbf{X_3}$ is vertically down; components of $\mathbf{B}$, for example, can alternatively be designated B_N, B_E, B_V.

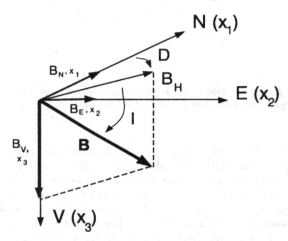

Figure 1.5. Components of the geomagnetic field vector $\mathbf{B}$. B_H is the projection of the field vector $\mathbf{B}$ onto a plane tangent to the Earth's surface. B_H can be resolved into north and east components (B_N and B_E). B_V is the projection onto the vertical axis. D is measured clockwise from North and ranges from $0 \to 360°$. I is measured positive down from the horizontal and ranges from $-90 \to +90°$ (because field lines can also point out of the Earth). $\mathbf{M}$ or $\mathbf{H}$ can be substituted for $\mathbf{B}$ as needed.

From Figure 1.5 we see how to convert from the angular coordinate system of declination, inclination and total field magnitude to cartesian coordinate systems, using a little trigonometry, *i.e.*,

$$B_H = B \cos I = \sqrt{B_E^2 + B_N^2} \quad \text{and} \quad B_V = B \sin I = x_3. \qquad (1.4)$$

The horizontal component can also be projected onto the north ($\mathbf{X_1}$) and East ($\mathbf{X_2}$) axes (the directions in which measurements are often made), *i.e.*,

$$B_N = x_1 = B \cos I \cos D \quad \text{and} \quad B_E = x_2 = B \cos I \sin D. \qquad (1.5)$$

Equations 1.4 and 1.5 work equally well for components of magnetization.
[See Example 1.1]

If you have the cartesian coordinates of **B** (or **H** or **M**), they can be transformed to the geomagnetic elements D, I and B:

$$\begin{aligned}
B &= \sqrt{x_1^2 + x_2^2 + x_3^2}, \\
D &= \tan^{-1}(x_2/x_1), \\
I &= \sin^{-1}(x_3/B).
\end{aligned} \qquad (1.6)$$

Be careful of the sign ambiguity of the tangent function. You may end up in the wrong quadrant and have to add 180°
[See Example 1.2]

1.3.2. PLOTTING MAGNETIC DIRECTIONAL DATA

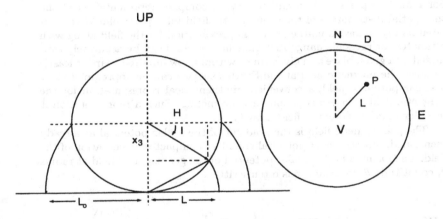

Figure 1.6. Construction of an equal area projection for a point P corresponding to a D of 40° and an I of 35°.

[See Example 1.3]

Magnetic field and magnetization directions can be visualized as unit vectors anchored at the center of a unit sphere. Such a unit sphere is difficult to represent on a 2-D page. There are several popular projections, including the Lambert equal area projection; we will be making extensive use of this projection in later chapters. The principles of construction of the equal area projection are shown in Figure 1.6. The point P corresponds to a D of 40° and I of 35°. D is measured around the perimeter of the equal area net and I is transformed as follows:

$$L = L_o\sqrt{(1 - |x_3|)},\qquad\qquad(1.7)$$

where $L_o = 1/\sqrt{x_1^2 + x_2^2}$.

In general, regions of equal area on the sphere project as equal area regions on this projection, as the name implies. Plotting directional data in this way enables rapid assessment of data scatter. A drawback of this projection is that circles on the surface of a sphere project as ellipses. Also, because we have projected a vector onto a unit sphere, we have lost information concerning the magnitude of the vector. Finally, lower and upper hemisphere projections must be distinguished with different symbols. The paleomagnetic convention is: lower hemisphere projections use solid symbols, while upper hemisphere projections are open.

1.3.3. REFERENCE MAGNETIC FIELD

For many purposes, it is useful to have a compact representation of the the spatial distribution of the geomagnetic field for a particular time. It is often handy to have a mathematical approximation for the field along with estimates for rates of change such that field vectors can be accurately estimated at a given place at a given time (within a few hundred years at least). Because the magnetic field at the Earth's surface can be approximated by a scalar potential field, a convenient mathematical representation for the magnetic field is in terms of spherical harmonics. Such a representation is used for another potential field, gravity.

The geomagnetic field is the gradient of the scalar potential as already mentioned, but the scalar potential is a more compact representation of the field. The formula for the scalar potential of the geomagnetic field at radius r, co-latitude θ, longitude ϕ is often written:

$$V(r,\theta,\phi) = a \sum_{l=1}^{\infty} \sum_{m=0}^{l} P_l^m(\cos\theta)\left[\left((g_e)_l^m\left(\frac{r}{a}\right)^l + (g_i)_l^m\left(\frac{a}{r}\right)^{l+1}\right)\cos m\phi\right.$$

$$\left. + \left((h_e)_l^m\left(\frac{r}{a}\right)^l + (h_i)_l^m\left(\frac{a}{r}\right)^{l+1}\right)\sin m\phi\right],\qquad(1.8)$$

where g and h are *Gauss coefficients* calculated for a particular year and are given in units of nT, or magnetic flux. The e and i subscripts indicate fields of external or internal origin and a is the radius of the Earth (6.371×10^6 m) and the P_l^m's are proportional to the Legendre polynomials, normalized according to the convention of Schmidt (see, for example, Chapman and Bartels [1940] and Backus et al. [1996]). The Schmidt polynomials are increasingly wiggly functions of the argument $\cos\theta$. Examples are:

$$P_1^0 = \cos\theta, \qquad P_2^0 = \frac{1}{4}(3\cos 2\theta + 1), \quad \text{and} \quad P_3^0 = \frac{1}{8}(5\cos 3\theta + 3\cos\theta).$$

Once the scalar potential is known, the components of the magnetic field can be calculated by the following relationships:

$$B_N = -\frac{1}{r}\frac{\partial V}{\partial\theta}, \quad B_E = -\frac{1}{r\sin\theta}\frac{\partial V}{\partial\phi}, \quad B_V = -\frac{\partial V}{\partial r}, \qquad (1.9)$$

where r, θ, ϕ are radius, co-latitude (degrees away from the North pole) and longitude, respectively. Here, B_V is positive down and to the north, the opposite of H_r and H_θ as defined in Figure 1.3. Note that equation 1.8 is in units of tesla, not Am^{-1} as in equation 1.3.

The Gauss coefficients are determined by fitting equations 1.9 and 1.8 to observations of the magnetic field made by magnetic observatories or satellite data for a particular time epoch.

The *International (or Definitive) Geomagnetic Reference Field* for a given time interval is an agreed upon set of values for a number of Gauss coefficients, and their time derivatives. The IGRF for 1995 is given in Table 1.2 (see Barton [1996]). IGRF (or DGRF) models and programs for calculating various components of the magnetic field are available on the Internet from the National Geophysical Data Center; the address is http://www.ngdc.noaa.gov. Using the values listed in, for example, Table 1.2 or any more recent model with the equations 1.8 and 1.9, we can calculate values of B, D and I at any location on Earth. Examples of maps made from such calculations using the IGRF for 1995 are shown in Figure 1.7. These maps demonstrate that the field is a complicated function of position on the surface of the Earth.

[See Example 1.4]

The intensity values in Figure 1.7a are in general highest at the poles ($\sim 60~\mu$T) and lowest near the equator ($\sim 30~\mu$T), but the contours are not straight lines parallel to latitude as they would be for a field generated strictly by a geocentric axial dipole such as that shown in Figure 1.4. Similarly, a GAD would produce lines of inclination that vary in a regular way from -90° to +90° at the poles, with 0° at the equator; the contours would parallel the lines of latitude. Although the general trend in inclination shown in Figure 1.7b is similar to this GAD model field, there is considerable structure to the lines, which again suggests that the field is not perfectly described by a geocentric bar magnet.

Finally, if the field were a GAD field, declination would be everywhere zero. This is clearly not the case, as is shown by the plots of declination

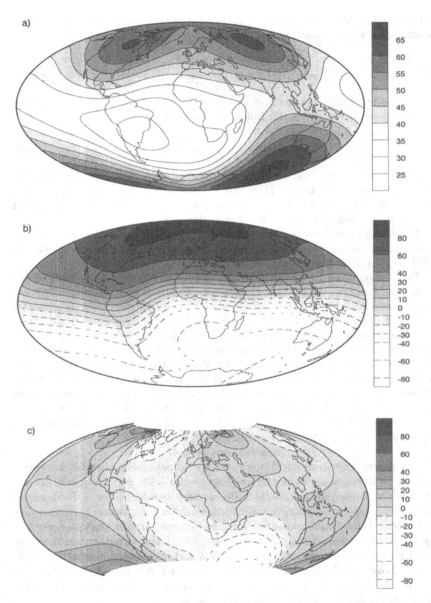

Figure 1.7.   Maps of geomagnetic elements of the IGRF for 1995: a) intensity in μT, b) inclination (in degrees), c) declination (in degrees).

TABLE 1.2. International Geomagnetic Reference Field, 1995.

m	l	Main field (nT) g	h	Change (nT yr^{-1}) $\dot{g}$	$\dot{h}$	m	l	Main field (nT) g	h	Change (nT yr^{-1}) $\dot{g}$	$\dot{h}$
0	1	-29682.0	0.0	17.60	0.00	2	6	65.0	77.0	0.60	-1.60
1	1	-1789.0	5318.0	13.00	-18.30	3	6	-172.0	67.0	1.90	-0.20
0	2	-2197.0	0.0	-13.20	0.00	4	6	2.0	-57.0	-0.20	-0.90
1	2	3074.0	-2356.0	3.70	-15.00	5	6	17.0	4.0	-0.20	1.00
2	2	1685.0	-425.0	-0.80	-8.80	6	6	-94.0	28.0	0.00	2.20
0	3	1329.0	0.0	1.50	0.00	0	7	78.0	0.0	-0.20	0.00
1	3	-2268.0	-263.0	-6.40	4.10	1	7	-67.0	-77.0	-0.80	0.80
2	3	1249.0	302.0	-0.20	2.20	2	7	1.0	-25.0	-0.60	0.20
3	3	769.0	-406.0	-8.10	-12.10	3	7	29.0	3.0	0.60	0.60
0	4	941.0	0.0	0.80	0.00	4	7	4.0	22.0	1.20	-0.40
1	4	782.0	262.0	0.90	1.80	5	7	8.0	16.0	0.10	0.00
2	4	291.0	-232.0	-6.90	1.20	6	7	10.0	-23.0	0.20	-0.30
3	4	-421.0	98.0	0.50	2.70	7	7	-2.0	-3.0	-0.60	0.00
4	4	116.0	-301.0	-4.60	-1.00	0	8	24.0	0.0	0.30	0.00
0	5	-210.0	0.0	0.80	0.00	1	8	4.0	12.0	-0.20	0.40
1	5	352.0	44.0	0.10	0.20	2	8	-1.0	-20.0	0.10	-0.20
2	5	237.0	157.0	-1.50	1.20	3	8	-9.0	7.0	0.40	0.20
3	5	-122.0	-152.0	-2.00	0.30	4	8	-14.0	-21.0	-1.10	0.70
4	5	-167.0	-64.0	-0.10	1.80	5	8	4.0	12.0	0.30	0.00
5	5	-26.0	99.0	2.30	0.90	6	8	5.0	10.0	0.20	-1.20
0	6	66.0	0.0	0.50	0.00	7	8	0.0	-17.0	-0.90	-0.70
1	6	64.0	-16.0	-0.40	0.30	8	8	-7.0	-10.0	-0.30	-0.60

in Figure 1.7c. Perhaps the most important result of spherical harmonic analysis for our purposes is that the field is dominated by the first order terms ($l = 1$) and the external contributions are very small. The first order terms can be thought of as geocentric dipoles that are aligned with three different axes: the spin axis (g_1^0) and two equatorial axes that intersect the equator at the Greenwich meridian (g_1^1) and at 90° East (h_1^1).

The vector sum of the geocentric dipoles is a dipole that is currently inclined by 11° to the spin axis. The axis of this so-called *best-fitting dipole* pierces the surface of the Earth at the diamond in Figure 1.8. This point and its antipode are called *geomagnetic poles*. Points at which the field is vertical ($I = \pm 90°$ shown by a triangle in Figure 1.8) are called *magnetic poles*, or sometimes, *dip poles*. These poles are distinguished from the *geographic poles* where the spin axis of the Earth intersects its surface. The Northern

Figure 1.8. The different magnetic poles. The triangle is the magnetic North Pole, where the magnetic field is straight down ($I = +90°$). The diamond is the geomagnetic North Pole, where the axis of the best fitting dipole pierces the surface. The dot is the geographic North Pole. The dashed line is the magnetic equator where $I = 0°$.

geographic pole is shown by a dot in Figure 1.8. Averaging ancient magnetic poles over some 10,000 years gives what is known as a *paleomagnetic pole*.

Because the geomagnetic field is axially dipolar to a first order approximation, we can write:

$$V = a g_1^0 \left(\frac{a}{r}\right)^2 P_1^0(\cos\theta) = a g_1^0 \left(\frac{a}{r}\right)^2 \cos\theta = \frac{B_o \cos\theta}{r^2}, \qquad (1.10)$$

where B_o is $g_1^0 a^3$ (g_1^0 can be read from Table 1.2).

Thus, from equation 1.10,

$$B_N = \frac{B_o \sin\theta}{r^3}, \qquad B_E = 0, \quad \text{and} \quad B_V = \frac{2 B_o \cos\theta}{r^3}. \qquad (1.11)$$

Consider some latitude λ on the surface of the Earth in Figure 1.4. Using the equations for B_V and B_N, we find that:

$$\tan I = \frac{B_V}{B_N} = 2\cot\theta = 2\tan\lambda. \qquad (1.12)$$

This equation is sometimes called the *dipole formula* or *dipole equation* which shows that the inclination of the magnetic field is directly related to the co-latitude for a field produced by a geocentric axial dipole (or g_1^0). This allows us to calculate the latitude of the measuring position from the inclination of the magnetic field, a result that is fundamental in plate tectonic reconstructions. The intensity of a dipolar magnetic field is also related to (co)latitude because:

$$B = (B_V^2 + B_N^2)^{\frac{1}{2}} = \frac{B_o}{r^3}(\sin^2\theta + 4\cos^2\theta)^{\frac{1}{2}} = \frac{B_o}{r^3}(1 + 3\cos^2\theta)^{\frac{1}{2}}. \quad (1.13)$$

The dipole field intensity has changed by more than an order of magnitude in the past and the dipole relationship of intensity to latitude turns out to be unuseful for tectonic reconstructions.

The dipole formula assumes that the magnetic field is exactly axial. Because there are more terms in the geomagnetic potential than just g_1^0, we know that this is not true. Because of the non-axial geocentric dipole terms, a given measurement of I will yield an equivalent *magnetic co-latitude* θ_m:

$$\cot\theta_m = \tfrac{1}{2}\tan I. \quad (1.14)$$

Paleomagnetists often assume that θ_m is a reasonable estimate of θ and the validity of this assumption depends on several factors. We consider first what would happen if we took random measurements of the Earth's present field (see Figure 1.9). We randomly selected 200 positions on the globe (shown in Figure 1.9a) and evaluated the direction of the magnetic field at each site using the IGRF for 1995. These directions are plotted in Figure 1.9b using the paleomagnetic convention of open symbols pointing up and closed symbols pointing down. We also plot the inclinations as a function of latitude on Figure 1.9c. We see that, as expected from a predominantly dipolar field, inclinations cluster around the values expected for a geocentric axial dipolar field.

1.3.4. VIRTUAL GEOMAGNETIC POLES

We are often interested in whether the geomagnetic pole has changed, or whether a particular piece of crust has rotated with respect to the geomagnetic pole. Yet, what we observe at a particular location is the local direction of the field vector. Thus, we need a way to transform an observed direction into the equivalent geomagnetic pole.

In order to remove the dependence of direction merely on position on the globe, we imagine a geocentric dipole which would give rise to the observed magnetic field direction at a given latitude (λ) and longitude (ϕ). The

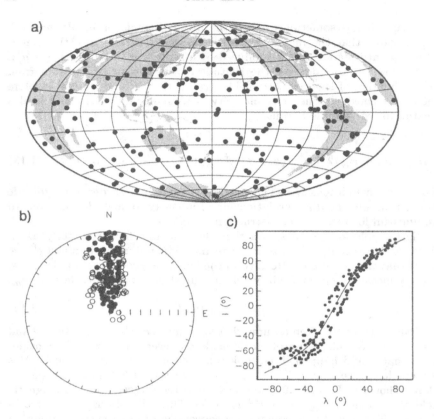

Figure 1.9. a) Hammer projection of 200 randomly selected locations around the globe. b) Equal area projection of directions of Earth's magnetic field as given by the IGRF evaluated for the year 1995 at locations shown in a). Open (closed) symbols indicate upper (lower) hemisphere. c) Inclinations (I) plotted as a function of site latitude (λ). The solid line is the inclination expected from the dipole formula (see text). Negative latitudes are south and negative inclinations are up.

virtual geomagnetic pole (VGP) is the point on the globe that corresponds to the geomagnetic pole of this imaginary dipole (Figure 1.10).

Paleomagnetists use the following conventions: ϕ is measured positive eastward from the Greenwich meridian and goes from $0 \rightarrow 360°$. θ is measured from the North pole and goes from $0 \rightarrow 180°$. Of course θ relates to latitude, λ by $\theta = 90 - \lambda$. θ_m is the magnetic co-latitude and is given by equation 1.14. Be sure not to confuse latitudes and co-latitudes. Also, be careful with declination. Declinations between 180 and 360° are equivalent to D - 360 and are counter-clockwise with respect to North.

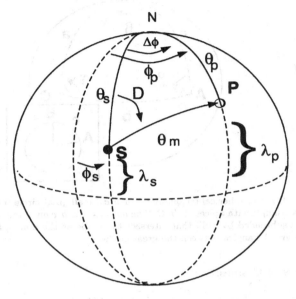

Figure 1.10.   Transformation of a direction measured at S into a virtual geomagnetic pole position P, using principles of spherical trigonometry and the dipole formula. Site S has latitude λ_s and longitude ϕ_s and a magnetic field direction D and I. The co-latitude of S is θ_s. $\Delta\phi$ is the difference in longitude between S and P and θ_p is the co-latitude of P. The direction can be transformed to an equivalent VGP (at P) with latitude λ_p and longitude ϕ_p as described in the text. The co-latitude of S with respect to P is the *magnetic co-latitude* θ_m. N is the geographic North Pole (the spin axis of the Earth).

[See Examples 1.5 and 1.6]

The first step in the problem of calculating a VGP is to determine the magnetic co-latitude θ_m by equation 1.14. Furthermore, the declination D is the angle from the geographic North Pole to the great circle joining S and P, and $\Delta\phi$ is the difference in longitudes between P and S, $\phi_p - \phi_s$. Now we need some tricks from spherical trigonometry.

In Figure 1.11, α, β and γ are the angles between the great circles labelled a, b, and c. On a unit sphere, a, b and c are also the angles subtended by radii that intersect the globe at the apices A, B, and C (see inset on Figure 1.11). Two formulae from spherical trigonometry come in handy for the purpose of calculating VGP, the Law of Sines:

$$\frac{\sin\alpha}{\sin a} = \frac{\sin\beta}{\sin b} = \frac{\sin\gamma}{\sin c}, \qquad (1.15)$$

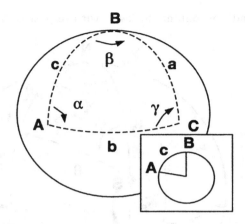

Figure 1.11. Rules of spherical trigonometry. a, b, c are all great circle tracks on a sphere which form a triangle with apices A, B, C. The lengths of a, b, c on a unit sphere are equal to the angles subtended by radii that intersect the globe at the apices, as shown in the inset. α, β, γ are the angles between the great circles.

and the Law of Cosines:

$$\cos a = \cos b \cos c + \sin b \sin c \cos \alpha. \tag{1.16}$$

We can locate VGPs using these two relationships. The declination D is the angle from the geographic North Pole to the great circle joining S and P (see Figure 1.10) so:

$$\cos \theta_p = \cos \theta_s \cos \theta_m + \sin \theta_s \sin \theta_m \cos D, \tag{1.17}$$

which allows us to calculate the VGP co-latitude θ_p. The VGP latitude is given by:

$$\lambda_p = 90 - \theta_p,$$

so $90 > \lambda_p > 0$ in the northern hemisphere and $0 < \lambda_p < 90$ in the southern hemisphere.

To determine ϕ_p, we first calculate the angular difference between the pole and site longitude $\Delta\phi$.

$$\sin \Delta\phi = \sin \theta_m \cdot \frac{\sin D}{\sin \theta_p}. \tag{1.18}$$

If $\cos \theta_m \geq \cos \theta_s \cos \theta_p$, then $\phi_p = \phi_s + \Delta\phi$. However, if $\cos \theta_m < \cos \theta_s \cos \theta_p$ then $\phi_p = \phi_s + 180 - \Delta\phi$.

Now we can convert the directions in Figure 1.9b to VGPs (Figure 1.12). The grouping of points is much tighter in Figure 1.12 than in the equal area projection because the effect of latitude variation in dipole fields has been removed.

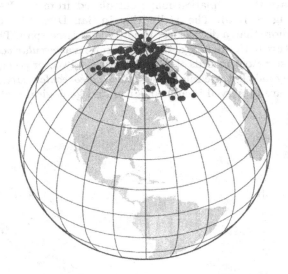

Figure 1.12.   VGP positions converted from directions shown in Figure 1.9b.

1.3.5.   VIRTUAL DIPOLE MOMENT

As pointed out earlier, magnetic intensity varies over the globe in a similar manner as inclination. It is often convenient to express paleointensity values in terms of the equivalent geocentric dipole moment which would have produced the observed intensity at that paleolatitude. Such an equivalent moment is called the *virtual dipole moment* (VDM) by analogy to the VGP. First, the magnetic paleoco-latitude θ_m is calculated as before from the observed inclination and the dipole formula of equation 1.12, then following the derivation of equation 1.13,

$$\text{VDM} = \frac{4\pi r^3}{\mu_o} B_{ancient}(1 + 3\cos^2\theta_m)^{-\frac{1}{2}}. \qquad (1.19)$$

Sometimes the site co-latitude as opposed to magnetic co-latitude is used in the above equation, giving a *virtual axial dipole moment* (VADM).

1.4. Earth's ancient magnetic field

1.4.1. PALEO-SECULAR VARIATION, EXCURSIONS AND REVERSALS

It is well known that magnetic field direction and intensity change with
time. Compare the declination maps calculated from the IGRF for 1945
and 1995 (Figure 1.13). The declination in San Diego, for example, has
changed by more than a degree over the fifty year time-span. The constantly
changing nature of the geomagnetic field is known as *secular variation* (SV).
There are observatory records of the magnetic field vector going back several
centuries. Beyond that, we rely on so-called *paleosecular variation* or PSV
records that are preserved in archeological and geological materials.

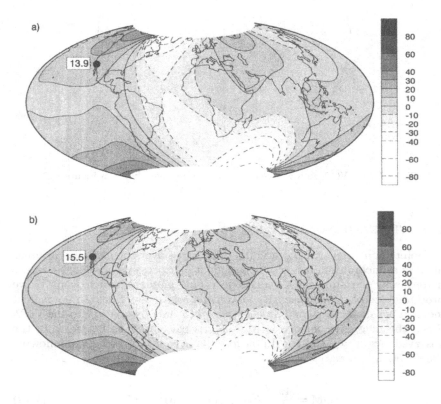

Figure 1.13.   Maps of declination evaluated for the IGRF of 1945 (a) and 1995 (b).

The restless nature of the non-dipole field appears to be an inherent part of the geodynamo process. In Figure 1.14, we see an example of what is observed at a single location over time. The geomagnetic field oscillates around the GAD direction with an amplitude of some 30° over an interval of some 9 meters (approximately 23 kyr) in the lake sediments that surround Mono Lake (Lund et al. [1988]). On rare occasions, the field departs drastically from what can be considered normal of secular variation and executes what is known as a *geomagnetic excursion.*

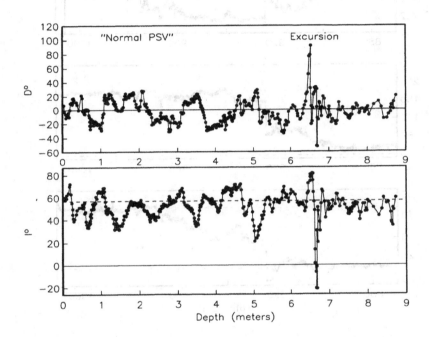

Figure 1.14.   Paleosecular variation of the magnetic field (*D* and *I*) observed at Mono Lake. The inclination expected from an geocentric axial dipole is shown as a dashed line. The declination is expected to be zero. The so-called "Mono Lake" excursion is marked. The data are from Lund et al. [1988] and represent some 23 kyr of time.

When viewed over sufficient time, the geomagnetic field reverses its polarity, by which we mean that the sign of the axial dipole changes. An example of a so-called *polarity reversal* is shown in Figure 1.15 (Clement and Kent [1984]). The intensity of the magnetic field appears to drop to approximately 10% of its average value and the directions migrate from one pole to the other over a period of approximately 4000 years.

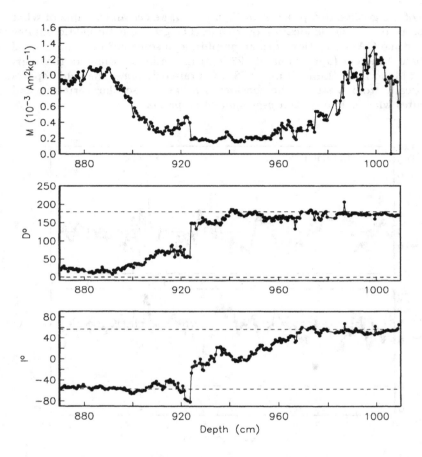

Figure 1.15. The lower Jaramillo geomagnetic polarity reversal as recorded in deep sea sediments from core RC14-14. Inclinations and declinations expected from a normal and reverse GAD field are shown as dashed lines. The data are from Clement and Kent [1984].

When the polarity is the same as the present polarity it is said to be *normal*. When it is in the opposite state, it is said to be *reverse*. On average, the field spends about half its time in each state, and only a tiny fraction (1-2%) of the time in an intermediate state. Rocks of both polarities have been documented from early in the Earth's history, although the frequency of reversal has changed considerably through time (see Opdyke and Channell

[1996] and Merrill et al. [1996]).

1.4.2. GEOMAGNETIC POLARITY TIME SCALE

A list of dates of past geomagnetic polarity reversals is known as a *geomagnetic polarity time scale* (GPTS). The first GPTS was calibrated for the last five million years by dating basalts of known polarity (see the excellent book by Glen [1982]). The polarity sequence is broken down into times of dominantly normal polarity and times of dominantly reverse polarity. These time units are known as *chrons*.

The uncertainty in the dating of young basalts exceeded the average duration of polarity intervals for times prior to about five million years until the advent of high precision ^{40}Ar/^{39}Ar dating techniques. The most complete historical record of paleomagnetic reversals (at least for the last 160 million years or so) is retained in the ocean crust. Modern time-scales are all based on the template provided by magnetic field anomalies measured by magnetometers towed across the oceans (see e.g., Cande and Kent [1992] and Figure 1.16).

Magnetic anomalies are generated at oceanic ridges or spreading centers, where molten rock from the mantle solidifies and acquires a magnetization during cooling (see Chapter 2). These strongly magnetized rocks are gradually carried away from the ridge by the process of seafloor spreading, and, as the polarity of the magnetic field changes, quasi-linear bands of oceanic crust with magnetizations of alternating polarity are generated. These bands produce magnetic fields that alternately add to and subtract from the Earth's ambient magnetic field, resulting in lineated magnetic anomalies. The anomalous magnetic field is obtained by subtracting the IGRF from the total magnetic field data (Figure 1.16c). These data are processed in order to make the anomalies as "square" as possible. Then, a square-wave is fitted to the data which are interpreted in terms of changes in polarity. In practice, several profiles can be stacked in order to average out noise and to produce a template that best represents the reversal history of the geomagnetic field.

The template of reversals obtained from marine magnetic anomaly data is in terms of kilometers from the ridge crest and covers the period of time for which there is sea floor remaining (since the Jurassic). The numerical calibration of the time-scale is frequently updated. All time-scale calibrations rely on the tying of numerical ages to known reversals (see e.g., Cande and Kent [1992] and Harland et al. [1989]). Ages for other reversals are interpolated or extrapolated. Numerical age information in recent time scales comes from both decay of radioactive isotopes and from calculations of long-term variations in the Earth's orbit (see e.g., Shackleton et al. [1990] and

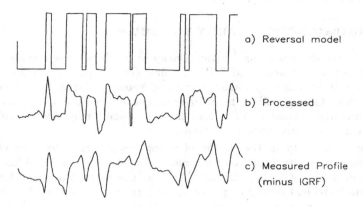

Figure 1.16.   a) The interpreted reversal model. b) Processed profile. c) Seasurface magnetic anomaly data over a portion of the East Pacific Rise. Redrawn from Cande and Kent [1992].

Hilgen [1991]). It is worth pointing out that the difference between the first Cenozoic time scale, based on extrapolation from about 3 Ma to about 80 Ma, and the most recent ones, which are based on many more dates, is no more than about 10%; the time scale is therefore reasonably well known and the current arguments are concerned with the details.

In Figure 1.17 we show the polarity history from the marine magnetic anomaly template (Cande and Kent [1992], Harland et al. [1989]). The details of the history of reversals for times older than the oldest sea floor magnetic anomaly record (about 160 Ma) are sketchy, but will eventually be documented using sedimentary records of the magnetic field (see Opdyke and Channell [1996]).

Examination of the reversal history shown in Figure 1.17 suggests that reversals occur at apparently random intervals without a predictable pattern. Furthermore, the frequency of reversals appears to change (see for example, Merrill et al. [1996]). Above the polarity pattern in Figure 1.17, we plot the number of reversals in four million year intervals as a histogram. The reversal frequency is relatively high in the interval 124-150 Ma, but appears to drop gradually to zero at the beginning of the so-called Cretaceous

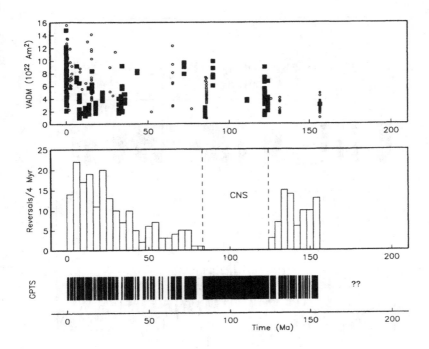

Figure 1.17. Bottom: The Geomagnetic Polarity Time Scale (GPTS) for the last 150 Ma (after Cande and Kent, [1992] and Harland et al. [1989]). Middle: Reversal frequency (number of reversals in a four million year interval). CNS is Cretaceous Normal Superchron. Top: paleointensity data from submarine basaltic glass (squares: data compiled by Juárez et al. [1998]; the dots are from the database of Tanaka et al. [1995]).

Normal Superchron (CNS), a period of some 40 m.y. in which no (or very few) reversals occurred. Since the end of the CNS at about 84 Ma, the frequency of reversals has increased to the present average rate of about four per million years.

1.4.3. PALEOINTENSITY

The magnetization in a rock, as well as retaining a record of the direction of the magnetic field when cooled from high temperature, has an intensity that is also a function of field magnitude. It is sometimes possible to estimate the magnitude of the Earth's field from geological samples (see e.g., Thellier and Thellier [1959] and Chapter 3). We plot compilations of such data

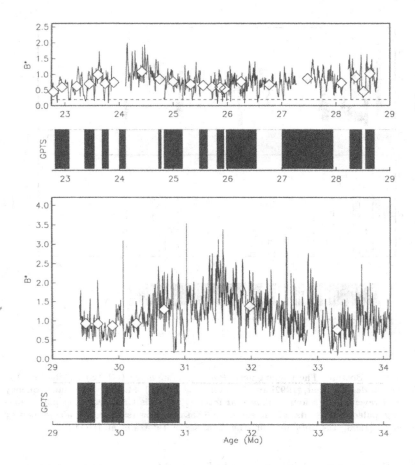

Figure 1.18.   Relative paleointensity data from sediments spanning the Oligocene (redrawn from Tauxe and Hartl [1996]). The data are normalized such that the average intensity of the entire record is unity. Average values from each polarity interval are plotted as diamonds. The time scale (GPTS) is that of Cande and Kent [1995]; periods of normal (reverse) fields are shown as black (white).

since the Jurassic in Figure 1.17, from Juárez et al. [1998] and Tanaka et al. [1995]. For the Tanaka et al. [1995] data, we plot only non-transitional data with more than a single sample normalized by the so-called Thellier-Thellier technique (see Chapter 3). Much of the Mesozoic had a rather low field intensity (the *Mesozoic dipole low* of Prévot et al. [1990]) with an apparent average intensity of about 25% of the present field which is ~ 8 x 10^{22} Am^2. The compilation of high quality paleointensity data by Juárez et al. [1998] shows that the Cenozoic also had a predominantly low field, suggesting that the Mesozoic "dipole low" is probably the most common state of the geomagnetic field, with anomalously high values occurring in the latter part of the Cretaceous and early Cenozoic and during the last few thousand years.

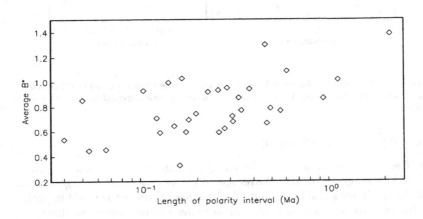

Figure 1.19.   Average paleointensity for polarity intervals (diamonds in Figure 1.18) plotted against polarity interval length. The data are redrawn from Tauxe and Hartl [1996].

The frequency of polarity reversals changes dramatically from the CNS to the present. The sparse paleointensity data in Figure 1.17 are consistent with the view that the long-term average field intensity has something to do with the average reversal frequency. The link between reversal frequency and paleointensity is more strongly made by the sedimentary paleointensity data of Tauxe and Hartl [1996] (Figure 1.18). These data indicate that the field is generally higher in the early part of the Oligocene when there are fewer reversals (about 1.6 Ma^{-1}) than in the last half of the Oligocene when there were more reversals (about 4 Ma^{-1}). If we replot the polarity interval averages (diamonds in Figure 1.18) against polarity interval length in Figure 1.19, there is a weak but significant correlation between polarity interval length and average intensity.

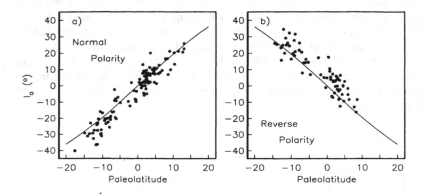

Figure 1.20. Observed inclination I_o versus paleolatitude from deep-sea sediment cores. Data of Schneider and Kent, [1990]. Solid line is the relation expected from the GAD hypothesis.

1.4.4. TIME-AVERAGED FIELD

It is often supposed that when averaged over some time interval, the directions of the magnetic field will average to those generated by a geocentric axial dipole. That is, all terms but g_1^0 will cancel. This is the GAD hypothesis. Because this hypothesis is central to many paleomagnetic applications, it is worthwhile considering its validity. If the GAD hypothesis is true, then the average declinations $\bar{D}$ of *in situ* rock formations should point directly toward the Earth's spin axis and the average inclinations $\bar{I}$ should relate to the site latitude λ by the dipole formula (equation 1.12).

In Figure 1.20, we plot inclination data from over 100 deep-sea sediment cores (Schneider and Kent [1990]) as a function of paleolatitude. The solid lines are the inclination expected from the dipole formula. It seems that the data fit the GAD model to first order, but data from reversed polarity intervals seem to be shifted to somewhat more positive values than expected from the GAD hypothesis (Schneider and Kent [1990]). The error in estimating paleolatitude that results from assuming a GAD model is about 3-4°. While this is not large, it should be kept in mind that the GAD hypothesis may not strictly be true (see also Johnson and Constable [1996]).

1.5. Examples

• Example 1.1

Use the program **dir_cart** to convert the following data from declination D, inclination I and intensity M to x_1, x_2, x_3.

D	I	M (μAm^2)
20	46	1.3
175	-24	4.2

Solution

The programs described in this book are listed in the Appendix, with instructions on their use. Also, all programs respond with help commands to the syntax: % **programname -h** where the % symbol stands for the command line prompt. So to find out what **dir_cart** does, type:

% **dir_cart -h**

to which the computer responds:

Usage: dir_cart [-m] [Standard I/O]

   options:

      -m read magnitude field

      input:

         declination, inclination, [magnitude]

      or

         longitude,latitude

      output:

         x1,x2,x3

From this, we see that to convert D, I, M data to cartesian components, we can type the following:

% **dir_cart -m**

20 46 1.3   [Input D, I, M]

  0.848595 0.308863 0.935142   [Output x_1, x_2, x_3]

175 -24 4.2

  -3.82229 0.334409 -1.70829

[type <control-D> to finish]

or:

enter D, I, M data into data file by typing:

% **cat>ex1.1**

20 46 1.3

175 -24 4.2

[type <control-D> to finish]

then type:

% **dir_cart -m < ex1.1**

and the computer responds:

  0.848595 0.308863 0.935142

  -3.82229 0.334409 -1.70829

Taking advantage of the UNIX ability to redirect output, the output can be put into a file by:

% **dir_cart -m** < **ex1.1** > **ex1.1a**

ex1.1a can be printed (with **lpr** on many systems), listed to the screen (with **cat** or **more**), or used as input for another program with the UNIX pipe facility.

• **Example 1.2**

Use the program **cart_dir** to convert these cartesian coordinates to geomagnetic elements:

x_1	x_2	x_3
0.3971	-0.1445	0.9063
-0.5722	.0400	-0.8192

Solution

Type:

% **cart_dir**

0.3971 -.1445 0.9063 [Input x_1, x_2, x_3]

  340.0 65.0 0.100E+01 [Output D, I, M]

-.5722 .0400 -.8192

  176.0 -55.0 0.100E+01

[type <control-D> to finish]

or use **cat** as in the first example.

• **Example 1.3**

Use the program **eqarea** to plot an equal area projection of the following directions.

D	I	D	I
346.5	2.7	334.4	-12.9
49.4	42.6	8.9	31.2
340.9	0.1	70.5	29.7
349.8	-12.4	166.3	20.7
169.7	2.3	182.3	15.0
196.4	39.3	165.9	3.6
186.1	2.5		

Solution
First enter the data **ex1.3** (using e.g., **cat**, as in Example 1.1). Then type:
% **eqarea** < **ex1.3** | **plotxy**
The program **plotxy** produces Postscript output which can be viewed with
ghostscript, ghostview, or any other Postscript viewer or manipulator.
The output should look like Figure 1.21.

North

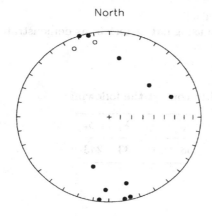

Figure 1.21. The file **mypost** generated with command **eqarea** < **ex1.3** | **plotxy** in
Example 1.3.

● **Example 1.4**
Use the program **igrf** to estimate the field on June 1, 1995 in Amsterdam,
The Netherlands (52.5°N, 5°E).
Solution
Type the following:
% **igrf**
1995.5 0 52.5 5 [Input decimal year, altitude (km), lat. (N), long. E)]
   358.0 67.4 48549 [Output D, I, B (nT)]
[type <control-D> to finish]
or use **cat** as in the first example.

● **Example 1.5**
Use the program **di_vgp** to convert the following:

D	I	λ_s (N)	ϕ_s (E)
11	63	55	13
154	-58	45.5	-73

Solution
Type the following:
% di_vgp
11 63 55 13 [Input D, I, λ_s, ϕ_s]
    154.7 77.3 [Output ϕ_p, λ_p]
154 -58 45.5 -73
    6.6 -69.6
[type <control-D> to finish]
or put the data into a file using **cat** > **ex1.5**, as demonstrated in Example
1.1. Then use the following:
% **di_vgp** < **ex1.5**

• **Example 1.6**
Use the program **vgp_di** to convert the following:

λ_p	ϕ_p	λ_s	ϕ_s
68	191	33	243

Solution
Type the following:
% **vgp_di**
68 191 33 243 [Input data]
    335.6 62.9 [Output answer]
[type <control-D> to finish]
or use **cat**, as illustrated in Example 1.1.

Chapter 2

RUDIMENTS OF ROCK MAGNETISM

Scientists in the late 19th century considered that it might be possible to exploit the magnetic record retained in rocks in order to study the geomagnetic field in the past. Early work in rock magnetism provided the theoretical and experimental basis for presuming that rocks might retain a record of past geomagnetic fields. There are several books and articles that describe the subject in detail (see e.g., Stacey and Banerjee [1974], O'Reilly [1984], Fuller [1987], Dunlop and Özdemir [1997]), while Butler [1992] gives an excellent introduction for the general earth science audience. We present here a brief overview of theories on how rocks become and stay magnetized.

Substances generally respond to magnetic fields; a few generate them. Therefore it is convenient to separate the magnetization of a material $\mathbf{M}$ into two contributions: that which exists only in the presence of an external magnetic field (induced magnetization) and that which exists in zero external magnetic field (remanent magnetization). As stated in Chapter 1, most of the magnetic behavior of solids results from electronic spin. Classical physics suggests that the moment generated by an orbiting electron is proportional to its angular momentum. Quantum physics tells us that the angular momentum must be quantized. The fundamental unit of magnetic moment of electrons is termed the *Bohr magneton* $\mathbf{m}_b$ and has a value of 9.27×10^{-24} Am^2.

The magnetic moments of electrons respond to externally applied magnetic fields, which creates an induced magnetization $\mathbf{M}_I$ that is observable outside the substance. $\mathbf{M}_I$ is a function of the applied field $\mathbf{H}$, *i.e.*,

$$\mathbf{M}_I = \chi \mathbf{H}. \tag{2.1}$$

From Chapter 1, we see that χ is the *magnetic susceptibility*. At its simplest, χ can be treated as a scalar and is referred to as the *bulk magnetic susceptibility* χ_b. In detail, magnetic susceptibility can be quite complicated. The relationship between induced magnetization and applied field can be affected by crystal shape, lattice structure, dislocation density, state of stress, etc., which give rise to possible anisotropy of the susceptibility. Furthermore, there are only a finite number of electronic moments within a given volume. When these are fully aligned, the magnetization reaches saturation. Thus, magnetic susceptibility is both anisotropic and non-linear with applied field. We will explore the origin of magnetic susceptibility only briefly here.

2.1. Induced magnetization

2.1.1. ROLE OF ELECTRONIC ORBITS

The orbit of an electron can be characterized as a moving charge with velocity v_e and charge q (see Figure 2.1 and Aharoni [1996]). What keeps the charge in orbit is the balance of the attractive force of the proton drawing the electron towards the nucleus and the "centripetal force" pushing the electon away from the nucleus. The attracive force between the proton (q_1) and the electron (q_2) is given by Coulomb's law:

$$F = \frac{kq_1q_2}{r^2},$$

where k is Boltzmann's constant, 1.381×10^{-23} JK^{-1}, and the "centripetal force" is given by:

$$F = \frac{m_e v_e^2}{r} = 2\pi m_e \omega^2 r,$$

where m_e and ω are the electronic mass and orbital frequency, respectively. Balancing these two competing forces and solving for ω gives a fundamental orbital frequency ω_o. The tiny current generated by the electronic orbit creates a magnetic moment. In the presence of an external field **H**, there is a torque $q_e v_e \times \mu_o H$ on the electron. The new balance of forces changes ω by some increment $\omega_L = -q_e v_e \mu_o H$, which is known as the *Larmor frequency*. The change in orbital frequency results in a changed magnetic moment. The change in net magnetization **M_I** is inversely proportional to **H**. The ratio **M_I/H** is the *diamagnetic susceptibility* χ_d; it is negative, essentially temperature independent, and quite small.

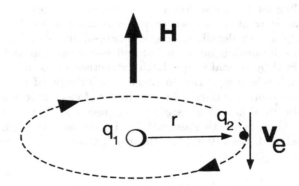

Figure 2.1. The balance of forces on an electron. The nucleus has a charge of q_1, while the electron has a charge of q_2, a velocity of v_e, and a mass of m_e.

2.1.2. ROLE OF ELECTRONIC SPINS

Unpaired electronic spins also behave as magnetic dipoles. In the absence of an applied field, or in the absence of the ordering influence of neighboring spins which are known as *exchange interactions*, the spins are essentially randomly oriented. An applied field acts to align the spins which creates a net magnetization equal to $\chi_p H$. χ_p is the *paramagnetic susceptibility*.

Each unpaired spin has a moment of one Bohr magneton $\mathbf{m}_b$. The elements with the most unpaired spins are the transition elements. These are responsible for most of the paramagnetic behavior observed in rocks.

A useful model for paramagnetism (see e.g., Chikazumi and Charap [1986] or Aharoni [1996]), Langevin theory, is based on a few simple premises:

- Each unpaired spin contributes a dipole moment.
- In the absence of an applied field, the moments are essentially randomly oriented, *i.e.*, all directions are equally likely to occur.
- An applied field acts to align the spins which creates a net moment.
- There is competition between thermal energy kT (T is temperature in kelvin) and the magnetic energy E_m. E_m of a magnetic moment $\mathbf{m}$ at an angle α with an external magnetic field $\mathbf{H}$ is given by:

$$E_m = -\mathbf{m} \cdot \mu_o \mathbf{H} = -m\mu_o H \cos\alpha, \tag{2.2}$$

where μ_o is the permeability of free space (see Table 1.1).

Magnetic energy is at a minimum when the magnetic moment is parallel to the magnetic field. Using the principles of statistical mechanics, we find that the probability density of a given moment having energy E_m is:

$$P(E) \propto \exp(-E_m/kT). \tag{2.3}$$

Because we have made the assumption that there is no preferred alignment within the substance, we can assume that the number of moments between angles α and $\alpha + d\alpha$ with respect to $\mathbf{H}$ is proportional to the solid angle $\sin\alpha\, d\alpha$ and the probability density function, *i.e.*,

$$n(\alpha)d\alpha \propto \exp\left(\frac{-E_m}{kT}\right)\sin\alpha\, d\alpha. \tag{2.4}$$

When we measure the induced magnetization, we really measure only the component of the moment parallel to the applied field (see Section 2.4 on hysteresis), or $mn(\alpha)\cos\alpha$. The net magnetization of a population of particles with volume v is therefore:

$$M_I = \frac{m}{v}\int_0^\pi n(\alpha)\cos\alpha\, d\alpha, \tag{2.5}$$

where $n(\alpha)$ is the number of moments between the angles α and $\alpha + d\alpha$; $n(\alpha)$ integrates to N, the total number of moments:

$$N = \int_0^\pi n(\alpha)d\alpha. \tag{2.6}$$

The total saturation moment of a given population of N individual magnetic moments m is Nm. The saturation value of magnetization M_s is thus Nm normalized by the volume v. Therefore, the magnetization expressed as the fraction of saturation is:

$$\frac{M}{M_s} = \frac{\int_0^\pi n(\alpha)\cos\alpha d\alpha}{\int_0^\pi n(\alpha)d\alpha}$$

$$= \frac{\int_o^\pi e^{(m\mu_o H \cos\alpha)/kT}\cos\alpha\sin\alpha d\alpha}{\int_o^\pi e^{(m\mu_o H \cos\alpha)/kT}\sin\alpha d\alpha}.$$

By substituting $a = m\mu_o H/kT$ and $\cos\alpha = x$, we write

$$\frac{M}{M_s} = N\frac{\int_{-1}^1 e^{ax}x dx}{\int_{-1}^1 e^{ax}dx} = (\frac{e^a + e^{-a}}{e^a - e^{-a}} - \frac{1}{a}), \tag{2.7}$$

and finally

$$\frac{M}{M_s} = [\coth a - \frac{1}{a}] = L(a). \tag{2.8}$$

The function enclosed in square brackets is known as the *Langevin function* $L(a)$ and is shown in Figure 2.2. It approaches saturation (in this case, M_s) when $m\mu_o H$ is some 10-20 times kT. When $kT >> m\mu_o H$, $L(a)$ is approximately linear with a slope of $\sim 1/3$. At room temperature and fields up to many tesla, L(a) is approximately $\mu_o m H/3kT$. If the moments m are unpaired spins ($m = m_b$), then $M_s = Nm_b/v$, and:

$$\frac{M}{M_s} \simeq \frac{m_b\mu_o}{3kT}H.$$

We have neglected all deviations from isotropy including quantum mechanical effects as well as crystal shape, lattice defects, and state of stress. We can rewrite the above equation as:

$$\frac{M}{H} = \frac{m_b\mu_o}{3kT} \cdot M_s = \frac{Nm_b^2\mu_o}{3kv} \cdot \frac{1}{T} = \chi_p. \tag{2.9}$$

To first order, paramagnetic susceptibility χ_p is: positive, larger than diamagnetism and inversely proportional to temperature. This inverse T dependence is known as Curie's law of paramagnetism.

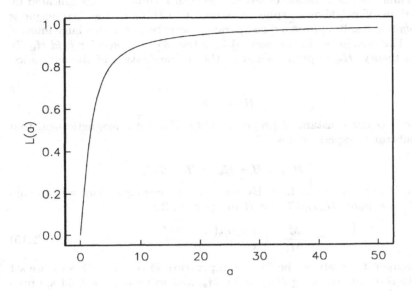

Figure 2.2. Langevin function $L(a)$ versus a.

2.2. Remanent magnetization

Some substances give rise to a magnetic field in the absence of an applied field. This magnetization is called *remanent* or *spontaneous* magnetization, and constitutes the phenomenon which is loosely known as *ferromagnetism (sensu lato)*. Magnetic remanence is caused by strong interactions between neighboring spins that occur in certain crystals. The so-called *exchange energy* is minimized when the spins are aligned parallel or anti-parallel depending on the details of the crystal structure. Exchange energy is a consequence of the quantum mechanical principle which states that no two electrons can have the same set of quantum numbers. In the transition elements, the **3d** orbital is particularly susceptible to exchange interactions because of its shape and the prevalence of unpaired spins, so remanence is characteristic of certain crystals containing transition elements with unfilled **3d** orbitals. As temperature increases, the scatter in spin directions also increases. Above a temperature characteristic of each crystal type (known as the *Curie temperature* Θ), cooperative spin behavior disappears and the material becomes paramagnetic.

While the phenomenon of ferromagnetism results from complicated interactions of neighboring spins, it is useful to think of the ferromagnetic moment as resulting from a quasi-paramagnetic response to a huge internal field. This imaginary field is termed here the *Weiss molecular field* H_w. In Weiss theory, H_w is proportional to the magnetization of the substance, *i.e.*,

$$H_w = \beta M,$$

where β is the constant of proportionality. The total magnetic field that the substance experiences is:

$$H_{tot} = H + H_w = H + \beta M,$$

where H is the external field. By analogy to paramagnetism, we can substitute $a = \mu_o m_b (H_{tot})/kT$ for H in equation 2.8

$$\frac{M}{M_s} = L\left(\frac{\mu_o m_b (H + \beta M)}{kT}\right). \tag{2.10}$$

For temperatures above the Curie temperature Θ (i.e. $T - \Theta > 0$), we set βM to zero. Substituting $N m_b/v$ for M_s, and using the low-field approximation for $L(a)$, equation 2.10 can be rearranged to get:

$$\frac{M}{H} = \frac{\mu_o N m_b^2}{v3k(T - \Theta)} \equiv \chi_f. \tag{2.11}$$

Equation 2.11 is known as the Curie-Weiss law and governs ferromagnetic susceptibility above the Curie temperature.

Below the Curie temperature, we can neglect the external field H and get:

$$\frac{M}{M_s} = L(\frac{\mu_o m_b \beta M}{kT}).$$

Substituting again for M_s and rearranging, we get:

$$\frac{M}{M_s} = L(\frac{N m_b^2 \beta}{vkT} \cdot \frac{M}{M_s}) = L(\frac{\Theta}{T} \cdot \frac{M}{M_s}), \tag{2.12}$$

where Θ is the Curie temperature and is given by:

$$\Theta = \frac{N m_b^2 \beta}{vk}.$$

Equation 2.12 can be solved graphically or numerically and is sketched in Figure 2.3. Below the Curie temperature, exhange interactions are strong

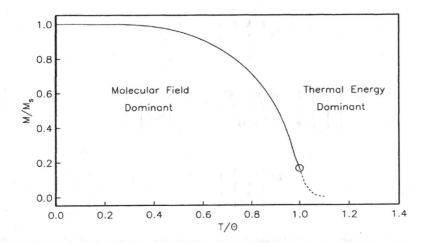

Figure 2.3. Behavior of magnetization versus temperature of a ferromagnetic substance.

relative to the external field and the magnetization is governed by equation 2.12. Above the Curie temperature, it follows the Curie-Weiss law (equation 2.11).

2.2.1. TYPES OF FERROMAGNETISM

As we have seen, below the Curie temperature, certain crystals have a permanent (remanent) magnetization resulting from the alignment of unpaired electronic spins over a large area within the crystal. Spins may be either parallel or anti-parallel; the sense of spin alignment is controlled entirely by crystal structure. The energy term associated with this phenomenon is the exchange energy. There are three categories of spin alignment: ferromagnetism (*sensu stricto*), ferrimagnetism and antiferromagnetism (see Figure 2.4).

In *ferromagnetism* (*sensu stricto*, Figure 2.4a), the exchange energy is minimized when all the spins are parallel, as occurs in pure iron. When spins are perfectly antiparallel (*antiferromagnetism*, Figure 2.4b), there is no net magnetic moment, as occurs in ilmenite. Occasionally, the antiferromagnetic spins are not perfectly aligned in an antiparallel orientation, but are canted by a few degrees. This *spin-canting* (Figure 2.4c) gives rise to a weak net moment, as occurs in hematite. Also, antiferromagnetic mate-

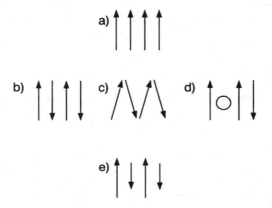

Figure 2.4. Types of spin alignment in ferromagnetism *(sensu lato)*: a) ferromagnetism *(sensu stricto)*, b) antiferromagnetism, c) spin-canted antiferromagnetism, d) defect anti-ferromagnetism, e) ferrimagnetism.

rials can have a net moment if spins are not perfectly compensated owing to defects in the crystal structure, as occurs in fine-grained hematite. The uncompensated spins result in a so-called *defect* moment (Figure 2.4d). Also, the temperature at which spins become disordered in antiferromagnetic substances is termed the *Néel temperature*. In *ferrimagnetism*, spins are also aligned antiparallel, but the magnitudes of the moments in each direction are unequal, resulting in a net moment (Figure 2.4e).

2.3. Magnetic anisotropy energy

Single crystals may have net magnetic moments which remain in the absence of an applied field. However, the direction of the net moment is free to rotate within the crystal, if it is not "blocked" by some other factor. Such a remanence would not have a long memory of ancient fields and would be useless for paleomagnetic purposes. The direction that a particular moment will have within a crystal will tend to lie in a direction that minimizes the magnetic energy. Magnetic anisotropy energy (see also O'Reilly [1984] and Dunlop and Özdemir [1997]) is responsible for blocking magnetic moments in particular directions within a crystal. By magnetic anisotropy, we mean that magnetic grains have "easy" directions of magnetization. A grain tends to be magnetized along these easy directions and energy is required to move the magnetic moment through the intervening "hard" directions.

An example of anisotropy energy resulting from crystal shape is illustrated in Figure 2.5.

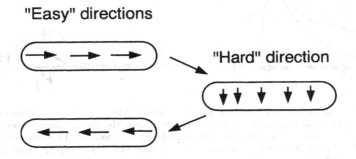

Figure 2.5. Illustration of one possible cause for preferred axes within magnetic crystals.

As anyone will remember from playing with magnets in school or on the refrigerator, magnetic moments prefer to be aligned head to tail, rather than with heads abutting. The magnetic energy of moments aligned along the length of a crystal is lower than those in which the moment is aligned crosswise. The direction along the length of the grain is "easy" and across the length is "hard". The energy required to move the moment from one easy direction to the other (through the hard direction) is the anisotropy energy. The case illustrated in Figure 2.5 is for anisotropy energy resulting from grain shape (magnetostatic energy). Other sources of anisotropy energy are crystal structure (magnetocrystalline energy) and the state of stress within the crystal (magnetostrictive energy). The magnitude of the magnetic field that supplies sufficient energy to overcome the anisotropy energy is called the *switching* or *coercive field* H_c.

Consider a particle with volume v whose easy axis makes an angle ϕ with the magnetic field **H** (see Figure 2.6). The magnetic moment **m** is drawn away from the easy axis, making an angle θ with the easy axis. The component of **m** parallel to **H** is given by $m_{||} = m \cos(\phi - \theta)$. The energy of such a particle is governed by two competing sources (Stoner and Wohlfarth [1948]). The anisotropy energy encourages alignment of **m** with the easy axis, and the magnetostatic energy $\mathbf{m} \cdot \mu_o \mathbf{H}$, acts to align **m** with **H**.

Anisotropy of the type depicted in Figure 2.5 and 2.6 has one preferred axis and a single constant K_u controls the magnetic anisotropy energy to first order. The anisotropy energy density for a uniaxial material is given by:

$$E_u = K_u \sin^2\theta. \tag{2.13}$$

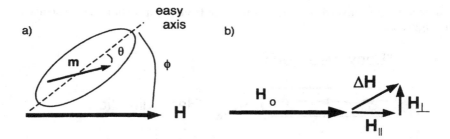

Figure 2.6. a) Sketch of a magnetic particle with easy axis as shown. In response to a magnetic field **H**, applied at an angle ϕ to the easy axis, the particle moment **m** rotates away from the easy axis, making an angle θ with the easy axis. b) The resultant field is $\mathbf{H}_o + \Delta\mathbf{H}$. $\Delta\mathbf{H}$ can be broken into components parallel ($\mathbf{H}_{\parallel}$) and perpendicular ($\mathbf{H}_{\perp}$) to $\mathbf{H}_o$. The contribution of magnetization induced by $\mathbf{H}_{\perp}$ can be neglected (see text).

Most often, uniaxial magnetic anisotropy results from anisotropic shape as in Figure 2.5. In this case, $K_u = \frac{1}{2}\Delta N M_s^2$, where ΔN is a dimensionless, shape dependent factor called the *demagnetizing factor*. This type of anisotropy energy is termed *magnetostatic energy* or *shape anisotropy*.

Magnetic anisotropy caused by magnetocrystalline sources can lead to several easy axes, depending on the symmetry of the crystal structure. In the case of a cubic crystal whose easy axis is aligned along the body diagonal (as in magnetite), the energy equation is somewhat more complicated than equation 2.13. One must take into account the relationship of **m** to the three crystal axes. Thus the magnetocrystalline anisotropy energy for a cubic mineral is:

$$E_c = K_1(\alpha_1^2\alpha_2^2 + \alpha_2^2\alpha_3^2 + \alpha_3^2\alpha_1^2) + K_2\alpha_1^2\alpha_2^2\alpha_3^2 \qquad (2.14)$$

where α_i are the direction cosines of **m** with respect to the crystal axes. K_1 and K_2 are controlled by exchange interactions between nearest neighbor electronic spins. In magnetite at room temperature, $K_1 < 0$ and $K_2 > 0$ so the easy axis is along the body diagonal ([111] direction). The magnetocrystalline anisotropy constants are strongly dependent on temperature, and changes in sign or relative magnitudes result in diagnostic features in thermomagnetic curves, as discussed later.

Another source of magnetic anisotropy is stress. Magnetic crystals change shape as a result of the ordering of magnetic moments below the Curie temperature, a phenomenon known as *magnetostriction*. As the spins move about in the crystal in response to applied fields, the crystal undergoes further deformation. The fractional change in length, dl/l is termed λ. Magnetostriction caused by a stress σ creates the anisotropy energy density E_σ, which is given to first order by:

$$E_\sigma = \frac{3}{2}\lambda\sigma\cos^2\psi, \qquad (2.15)$$

where ψ is the angle between **m** and the principal stress direction.

2.4. Magnetic hysteresis

Because the response of a magnetic substance to an applied field depends strongly on the physical properties of the material, it is rapidly becoming routine to measure what is known as *hysteresis loops* in rock and paleomagnetic studies (see e.g., Tauxe et al. [1996b] for a more complete discussion). While the interpretation of these loops is not simple, much can be learned in a short amount of time by analyzing the loops in an informed way.

Hysteresis loops are generated by subjecting a small sample to a large magnetic field H_{max}. The magnetization is monitored as the applied field decays to zero, switches polarity and approaches $-H_{max}$, then returns through zero to $+H_{max}$.

Before describing the analysis of loops in detail, we must first consider what controls the shape of loops in a simple system. If we imagine a particle similar to that illustrated in Figure 2.6 that has a single easy axis at some angle ϕ to the applied field, the direction of magnetization will be at some angle θ with respect to the easy axis, reflecting the balance of anisotropy and magnetic energies.

In the uniaxial case E_u is given by equation 2.13. The magnetic energy from the external field is $-\mathbf{m}\cdot\mu_o\mathbf{H}$. Thus, the energy density of the magnetic grain depicted in Figure 2.6 is given by:

$$E = K_u\sin^2\theta - M_s\mu_o H\cos(\phi - \theta). \qquad (2.16)$$

As shown in Figure 2.6b, the magnetometer is only sensitive to the induced component of **m** parallel to the applied field $\mathbf{H}_o$, which is $m_\parallel$.

If the easy axis is aligned parallel to the field ($\phi = 0$), the induced moment remains parallel to **H**. The angle $\theta = 0$ and the component of **m** parallel to **H** ($m_\parallel$) equals the magnitude of **m**, m (see the square loop in Figure 2.7). As the field decreases to zero and then to $-H_{max}$, $m_\parallel$ remains unchanged until the field is sufficiently large to cause **m** to switch through the intervening hard direction to the other direction along the easy axis (contributing $-m$ to the resulting loop). This field is known as the switching field and is a function of K_u, M_s, and ϕ. For $\phi = 0$, we define the switching field to be the intrinsic coercivity H'_c, which is related to K_u and M_s by:

$$H'_c = \frac{2K_u}{\mu_o M_s}.$$

Following the $\phi = 0$ curve from $-H'_c$ to $-H_{\max}$ and then to $H = 0$ in Figure 2.7, we have no change in $m_{||}$. On its final approach to $+H_{\max}$ the moment again switches at $+H'_c$ and then $m_{||}/m = 1$.

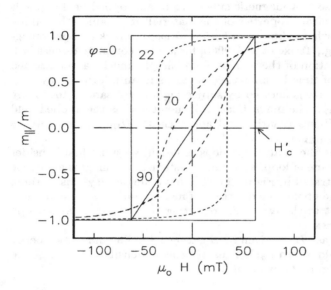

Figure 2.7. The component of **m** parallel ($m_{||}$) to **H** for various values of ϕ. The intrinsic coercive field is marked by H'_c.

The calculation of $m_{||}$ for ϕ other than 0 is more difficult. The portion of the loop from $H_{\max}$ to 0 (the "descending loop") is calculated by first numerically evaluating the smallest θ for which E is at a minimum. Then $m_{||} = m \cos(\phi - \theta)$. The portion back to $H_{\max}$ (the "ascending loop") must be calculated in two parts. We begin by assuming that the moments are in a state of saturation from exposure to $-H_{\max}$, and evaluate $m_{||}$ as a function of H for increasing H until the intrinsic coercive field is reached. At this point the ascending loop joins the descending loop. In Figure 2.7, we show several examples of loops for various values of ϕ.

In the case of $\phi = 90°$, the "loop" is a line. The moment is entirely "bent" into the direction of **H** for large **H**. As the field decreases, the moment relaxes back into the easy direction and $m_{\parallel}$ is zero at zero field.

In rocks, there are many individual magnetic particles whose contributions sum to produce the observed hysteresis loops. In Figure 2.8, we show the sum of 10,000 hypothetical grains with randomly oriented ϕ, whose magnetic anisotropy is uniaxial. After exposure to fields in excess of the intrinsic coercivities of all the grains in the sample (the *saturating field* H_s), a sample will have a *saturation remanent magnetization* M_r. In the uniaxial case, M_r/M_s is 0.5.

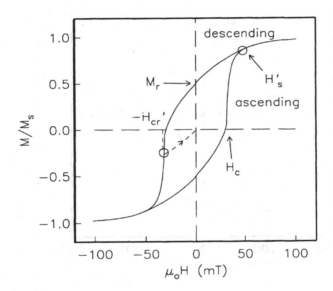

Figure 2.8. Sum of 10,000 hysteresis curves for individual grains with ϕ drawn from a uniform distribution on a sphere. The saturation remanence M_r, bulk coercivity H_c, coercivity of remanence H'_{cr}, and saturating field H'_s are indicated.

In some geologically important cases, minerals may have an anisotropy energy that is controlled by magnetocrystalline sources; hence they may not be uniaxial, but instead may be, say, cubic in symmetry (see e.g., Gee and Kent [1995]). In magnetite (at room temperature), the easy axis is along the body diagonal. There are four easy axes within a given crystal and the maximum angle ϕ between an easy axis and any **H** direction is 55°. Because this is much less than 90°, there is no individual loop that goes through the origin. A random assemblage of particles with cubic anisotropy will have a much higher saturation remanence. In Figure 2.9 we show a synthetic loop generated from 500 grains with randomly distributed crystal axes for which K_1 was that of magnetite (see Table 2.1) and the anisotropy energy was cubic in origin. The theoretical ratio of M_r/M_s for such assemblages is 0.87, as opposed to 0.5 for uniaxial anisotropy (see Joffe and Heuberger [1974]).

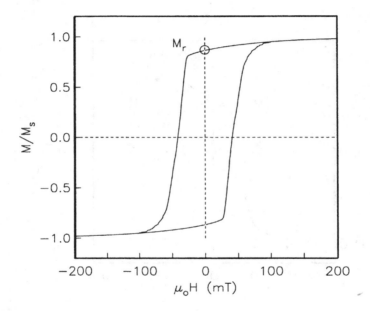

Figure 2.9. Sum of 500 individual hysteresis curves calculated for randomly oriented "grains" with cubic anisotropy. The theoretical value of M_r/M_s is 0.87 for such a population (see text). The value calculated here is 0.865.

2.4.1. HYSTERESIS PARAMETER ESTIMATION

Hysteresis loops are rich in information, but require a large number of parameters to describe them (a typical loop can consist of hundreds of measurements). It would be useful to characterize the main features of the loop with a few well chosen parameters. For example, the maximum magnetization achieved is the *saturation magnetization* M_s. In order to estimate this, data must be carefully adjusted to remove any para- or diamagnetic contributions. The maximum (saturation) remanence M_r acquired is estimated by the value at the y intercept of the descending curve (see Figure 2.8). The bulk (or average) coercivity H_c is the x intercept of the ascending loop. Another useful parameter is the field required to achieve the saturation remanence, the saturating field H_s. In a saturation hysteresis loop, the field required to "close" the loop whereby the ascending and descending curves join, is H_s and is indicated by H_s' in Figure 2.8. The prime indicates that this value was deterimined from a hysteresis loop, as opposed to direct measurement as described later.

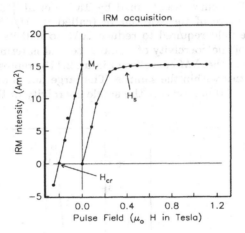

Figure 2.10. Acquisition of isothermal remanence after application of successive fields.

In Figure 2.10 we illustrate the acquisition of remanence after subjecting a sample to successively higher DC fields. This remanence is termed an *isothermal remanent magnetization* (IRM). In this case, H_s is around 0.4 tesla. The saturation IRM is the same as M_r calculated from hysteresis loops, as shown in Figure 2.8.

The value of the field necessary to remagnetize half the moments aligned at M_r, thereby reducing the net magnetization to 0, is termed the *coercivity of remanence* or H_{cr}. There are several ways of estimating H_{cr}. It can be measured directly by subjecting the saturation remanence to an increasingly strong field in the opposite direction and measuring remanence M_r

after each step. The field required to reduce M_r to zero is H_{cr} as illustrated in Figure 2.10. Alternatively, it can be calculated from the hysteresis loop, as shown in Figure 2.8. Imagine that, instead of continuing from 0 to the maximum field on the ascending loop, one switches the field off at some value of H larger than H_c, and allows the magnetization to relax back to some remanent value. The value of H which results in a net remanence of 0 provides another estimate of the coercivity of remanence which we call H'_{cr}. This parameter can be estimated numerically by sliding the descending loop down by the value of M_r (so that it has a zero intercept) and determining the field at which the x axis intercepts the ascending loop of the adjusted curve. This method gives is a reasonable estimate of H'_{cr}.

A third way to estimate H_{cr} is illustrated in Figure 2.11. The difference between the descending and ascending loops in Figure 2.8 (ΔM) is plotted in Figure 2.11a. This curve was termed by Tauxe et al. [1996b] the "ΔM curve". The derivative of the ΔM curve (called $d\Delta M/dH$) is shown in Figure 2.11b. The field required to reduce ΔM to half its initial value is another measure of the coercivity of remanence and is termed here H''_{cr}.

The derivative of the ΔM curve (see Figure 2.11b) represents the distribution of coercivities within the sample. The large hump in $\mu_o H$ centered on approximately $\sim$30 mT reflects the single coercivity of the population.

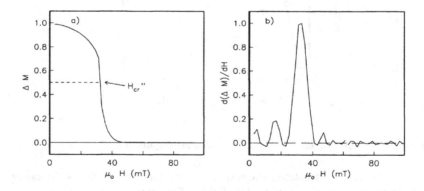

Figure 2.11.   a) The ΔM curve for the data shown in Figure 2.8. b) The derivative of the ΔM curve.

2.5. Magnetic domains

Until now, we have assumed that a magnetic crystal behaves as a single isolated magnetic dipole. Such grains are termed *single domain* (SD) grains.

In nature, this condition is rarely met. The "free poles" at the grain's surface create a magnetic energy which increases with grain volume. At some size, it becomes energetically more favorable to break the magnetization into several uniformly magnetized regions, or *magnetic domains,* as this reduces the associated magnetic field. Magnetic domains are separated by *domain walls.* Such grains are termed *multi-domain* (MD) grains.

Magnetic grains with few domains behave much like single domain grains in terms of magnetic stability and saturation remanence. These grains have earned the name *pseudo-single domain* (PSD) grains (*e.g.,* Stacey and Banerjee [1974]) and are apparently responsible for most of the stable remanence coveted by paleomagnetists.

The field produced by MD grains could be reduced in several ways (Figure 2.12). Each configuration has a penalty with respect to one or more of the various energy terms. For example, the circular spin option (Figure 2.12d), while eliminating the associated magnetic field, dramatically increases exchange energy.

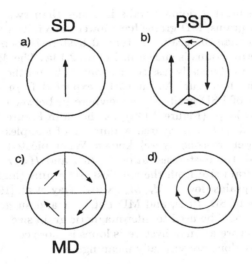

Figure 2.12. Possible domain structures: a) single domain, b) pseudo-single domain, c) multi-domain, d) circular spin state.

It requires a great deal of energy for magnetic grains to nucleate domain walls. Within the wall, the spins must change from one easy direction to another (see Figure 2.13). The narrower the wall, the greater the exchange energy because the spins are not parallel. The wider the wall, the greater the magnetocrystalline energy, because the spins will orient at some angle to the easy direction. The number of walls in a given grain will depend on its size, distribution of defects, state of stress and shape, to mention a few factors. The reader is referred to Dunlop and Özdemir [1997] for a more complete discussion of domain theories and observations.

Domain Wall

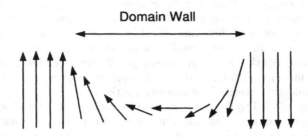

Figure 2.13. Illustration of the rotation of neighboring spins within a 180° domain wall.

Because moving domain walls is easier than switching the entire moment of SD grains, MD grains have lower coercivities and lower saturation remanences than SD grains. A typical hysteresis loop for a population of MD magnetite grains is shown in Figure 2.14a. The M_r/M_s ratio for such a population is typically less than about 0.05 and the H_{cr}/H_c ratio is typically larger than four. As might be expected from the name, loops for populations of PSD grains lie somewhere in between the SD (Figure 2.8) and the MD loops (Figure 2.14a), as shown in Figure 2.14b.

Day et al. [1977] measured a number of samples of magnetite whose grain sizes were reasonably well known. When plotted on a log-log plot (see Figure 2.15), the hysteresis ratios M_r/M_s and H_{cr}/H_c fall along a line of increasing grain size from the smallest SD grains (highest M_r/M_s) to the largest MD grains (lowest M_r/M_s ratios). Day et al. [1977] divided the plot into regions of SD, PSD, and MD behavior and suggested that hysteresis parameters could be used to infer magnetic grain size. As we shall see later, physical interpretation of hysteresis loops is more complex and simple plots of the ratios alone are virtually meaningless.

2.6. Mechanisms of remanence acquisition

We turn now to the subject of how geological materials become magnetized in nature. In order for rocks to stay magnetized in a particular direction, we must consider the role of the competition between exchange energy, anisotropy, and thermal energies in particles. There are a number of mechanisms by which magnetic remanence can be acquired. These are important to paleomagnetists because the mode of magnetic remanence acquisition is critical

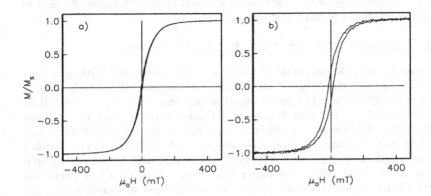

Figure 2.14. a) Typical loop for a population of MD grains. The data are from a gabbro from the Troodos Ophiolite (courtesy of J. Gee). b) Typical loop for a population of PSD grains. The data are from a marine carbonate.

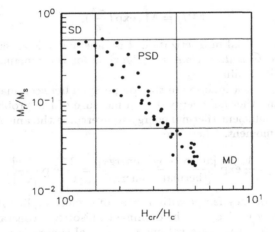

Figure 2.15. Hysteresis ratios from sized magnetite samples plotted on a log-log scale. Empirically determined boundaries of SD, PSD, and MD behavior are indicated. The data are from Day et al. [1977].

to the interpretation as to age and reliability of the remanence isolated in the laboratory. In the following, we will discuss the most important of these in some detail. Appendix 2 contains a list and short definition of many of the common forms of magnetic remanence.

2.6.1. MAGNETIC VISCOSITY AND RELAXATION TIME

The essence of paleomagnetic stability can be illustrated with a discussion of *magnetic viscosity*, or the change in magnetization with time at constant temperature. The following ideas are explained in more detail by Néel [1949, 1955] (see also Stacey and Banerjee [1974] and Dunlop and Özdemir [1997]). Imagine a block containing an assemblage of randomly oriented, non-interacting, uniformly magnetized particles. Let us further suppose that each particle has a single easy axis and that the magnetization lies in either direction along that axis. Occasionally, a particular particle has sufficient thermal energy to overcome the magnetic anisotropy energy associated with the intervening hard directions and the moment switches its direction along the easy axis. In the absence of an applied field, the moments of an assemblage of particles will tend to become randomly oriented and any initial magnetization will decay away according to the following equation:

$$\mathbf{M}(t) = \mathbf{M_o} \exp\left(\frac{-t}{\tau}\right), \tag{2.17}$$

where M_o is the initial magnetization, t is time and τ is an empirical constant called the *relaxation time*. τ is the time for the remanence to decay to $1/e$ of its initial value.

The value of τ is a function of the competition between magnetic anisotropy energy and thermal energy. It is a measure of the probability that a grain will have sufficient thermal energy to overcome the anisotropy energy and switch its moment. Therefore:

$$\tau = \frac{1}{C} \exp \frac{[\text{anisotropy} \quad \text{energy}]}{[\text{thermal} \quad \text{energy}]} = \frac{1}{C} \exp \frac{[Kv]}{[kT]}, \tag{2.18}$$

where C is a frequency factor with a value of something like 10^{10} s^{-1}. The anisotropy energy is given by the dominant anisotropy constant K (either K_u, K_1, or λ) times the grain volume v. Thermal energy is given by kT.

Thus, the relaxation time is proportional to coercivity, and volume, and is inversely related to temperature. Relaxation time τ varies rapidly with small changes in v and T. There is a sharp transition between grains with virtually no stability (τ is on the order of seconds) and grains with stabilities of 10^9 years. Grains with $\tau \simeq 10^2 - 10^3$ seconds have sufficient thermal energy to overcome the anisotropy energy frequently and are unstable on a laboratory time-scale. In zero field, these grain moments will tend to rapidly become random and in an applied field, they tend to rapidly align with the field. The net magnetization is related to the field by a Langevin function. Therefore, this behavior is quite similar to paramagnetism, hence these grains are called *superparamagnetic* (SP). Such grains can be distinguished from paramagnets, however, because the field required to saturate the moments is typically less than a tesla, whereas that for paramagnets can exceed hundreds of tesla.

2.6.2. VISCOUS REMANENT MAGNETIZATION

The magnetization which is acquired by viscous processes is called a *viscous remanent magnetization* or VRM. With time, more and more grains will have sufficient thermal energy to overcome anisotropy energy barriers and will switch their magnetizations to an angle that is more in alignment with the external field. If a specimen with zero initial remanence is put into a magnetic field, the magnetization $M(t)$ will grow to the *equilibrium magnetization* M_{eq} by the complement of equation 2.17:

$$M(t) = M_{eq}(1 - e^{-t/\tau}). \tag{2.19}$$

The more general case, in which the initial magnetization of a specimen is non-zero, can be written as (see Kok and Tauxe [1996a]):

$$M(t) = M_o + (M_{eq} - M_o)(1 - e^{-t/\tau}) = M_{eq} + (M_o - M_{eq}) \cdot e^{-t/\tau}, \tag{2.20}$$

which grows (or decays) exponentially from $M_o \to M_{eq}$ as $t \to \infty$. The rate is not only controlled by τ, but also by the degree to which the magnetization is out of equilibrium. Some data sets appear to follow the relation $M(t) \propto \log(t)$ (see e.g. Shimizu [1960]). Such a relation suggests infinite remanence as $t \to \infty$, which cannot be true over a long period of time. Such behavior can generally only be observed over a restricted time interval, long-term observations rarely show a strict $\log(t)$-behavior.

2.6.3. THERMAL REMANENT MAGNETIZATION

From equation 2.18 we know that τ is strongly temperature dependent. According to Néel's theory for single domain thermal remanence (Néel [1949, 1955]), there is a sharply defined range of temperatures over which τ increases from geologically short to geologically long time-scales. The temperature at which τ is equal to about $10^2 - 10^3$ seconds is defined as the *blocking temperature* (T_b). At or above the blocking temperature, but below the Curie temperature, a grain will be superparamagnetic. Further cooling increases τ such that the magnetization is effectively blocked and the rock acquires a geologically significant *thermal remanent magnetization* or TRM.

Consider a lava flow which has just been extruded. First, the molten lava solidifies into rock. While the rock is above the Curie temperature, there is no remanent magnetization; thermal energy dominates the system. As the rock cools through the Curie temperature of its magnetic phase, exchange energy becomes more important and the rock acquires a remanence. In the superparamagnetic state, the magnetization is free to track the prevailing magnetic field because magnetic anisotropy energy is still less important than the thermal energy.

The magnetic moments in the lava flow tend to flop from one easy direction to another, with a slight statistical bias toward the direction with the minimum angle to the applied field. Thus, the equilibrium magnetization of superparamagnetic grains is only slightly aligned, and the degree of alignment is a linear function of the applied field for low fields such as the Earth's. The magnetization approaches saturation at higher fields (from $\sim$ 0.2 T to several tesla, depending on the details of the source of anisotropy energy).

Now imagine that the lava continues to cool. The thermal energy will decrease until the magnetic anisotropy energy becomes important enough to "freeze in" the magnetic moment wherever it happens to be. As the particles cool through their blocking temperatures, the magnetic moments become fixed because τ reaches a time that is geologically meaningful.

From the preceeding discussion, we can make several predictions about the behavior of a TRM.

1. The remanence of an assemblage of randomly oriented particles, acquired by cooling through the blocking temperature in the presence of a field, should be parallel to the orientation of that field.

2. The intensity of thermal remanence should be linearly related to the intensity of the magnetic field applied during cooling (for weak fields such as the Earth's).

In a rock, each grain has its own blocking temperature and moment. Therefore, by cooling a rock between two temperatures, only a portion of the grains will be blocked; the rock thus acquires a *partial thermal remanent magnetization* or pTRM. An essential assumption in paleomagnetic applications is that each pTRM is independent of all others and that a pTRM acquired by cooling through two temperatures can be removed by exposure to the same peak temperature and cooling in zero field.

Experimental results tend to substantiate the pTRM theory outlined above although the behavior of non-SD grains appears to be quite different. For details on multi-domain TRM see Dunlop and Xu [1994], Xu and Dunlop [1995], and Dunlop and Özdemir [1997].

2.6.4. ANHYSTERETIC REMANENT MAGNETIZATION

Although not a naturally occurring remanence, it is worthwhile at this point to introduce a type of remanence that is closely analogous to TRM, but which is acquired in gradually declining oscillating magnetic fields instead of during cooling. Examination of equation 2.18 reveals that τ is dependent on the magnetic anisotropy energy of the grain. If, instead of raising the temperature, we subject a grain to an alternating field sufficient to overcome the anisotropy energy, the magnetization of the grain will follow the field. If we have a population of grains with a range of coercivities and we lower the peak field reached in each successive oscillation, the magnetic moments will get stuck in whatever direction they were pointing when the field went below their coercive fields. In zero field, the net magnetization will be zero. If there is a small DC bias field, then there will be a statistical preference for the direction of the bias field, which is analogous to the aquisition of TRM acquired during cooling. This net magnetization is termed the *anhysteretic remanent magnetization* or ARM. A *partial ARM* (pARM) is an ARM acquired when the DC field is applied between two specific values of the AF as opposed to the entire range from a saturating AF field to zero.

2.6.5. CHEMICAL REMANENT MAGNETIZATION

Inspection of equation 2.18 reveals that τ is also a strong function of grain volume. A similar theoretical framework can be built for remanence acquired by grains growing in a magnetic field as for those cooling in a magnetic field. As a starting point for our treatment, consider a diamagnetic porous matrix, say a sandstone. As ground water percolates through the sandstone, it begins to precipitate tiny grains of a magnetic mineral. Each new crystal is completely isolated from its neighbors. For very small grains, the thermal energy dominates the system and they are superparamagnetic. When the volume becomes sufficient for magnetic anisotropy energy to overcome the thermal energy, the grain moment is blocked and can remain out of equilibrium with any changes in the magnetic field for geologically significant time periods. Keeping temperature constant, there is a critical *blocking volume* (v_b) below which a grain maintains equilibrium with the applied field and above which it does not. Thus, the magnetization acquired during grain growth is controlled by the alignment of grain moments at the time that they grow through the blocking volume. Based on these principles, the so-called *chemical remanent magnetization* (CRM) should behave very similarly to TRM (see e.g., Haigh [1958]).

There have been a few experiments carried out with an eye to testing the CRM model, and although the theory predicts the zero-th order results quite well (that a simple CRM parallels the field and is proportional to it in intensity), the details are not well explained primarily because the magnetic field itself affects the growth of magnetic crystals and the results are not exactly analogous to TRM conditions (see e.g., Stokking and Tauxe [1990]).

CRM can also be acquired when one magnetic mineral alters into another. The behavior of such an alteration CRM is extremely complicated. It can parallel the magnetic field during alteration, or it can remain in the initial magnetization direction, or it can carry a magnetization with no discernable relationship to any past field (see Dunlop and Özdemir [1997] for a more complete discussion). In order to determine the reliability of magnetic remanence in general, we rely on a variety of techniques as described in later chapters.

2.6.6. DETRITAL REMANENT MAGNETIZATION

In sedimentary environments, rocks become magnetized in quite a different manner than igneous bodies (see Tauxe [1993] for a review). Detrital grains are already magnetized, unlike igneous rocks which crystallize from above their Curie temperatures. In the water column, where viscosity is low, there is a strong tendency for magnetic grains to become aligned with the magnetic field in response to the magnetic torque $\mathbf{m} \times \mu_o\mathbf{H}$. Nonetheless, there is competition among forces arising from turbulent motions of the water and perhaps from thermal agitation (Brownian motion) of the grains themselves. There is a small net alignment of magnetic grains in the direction of the prevailing field. When the grain strikes the sediment-water interface, this net moment may be disrupted and even distorted by gravitational effects that act on grains as they settle on the bottom. The remanence acquired is a *depositional* or *detrital remanent magnetization* or DRM.

Magnetic grains can be remobilized for some time after deposition, however, owing to the bioturbation of superficial sediments. If resuspended, the grains may realign with the field, thus acquiring a *post-depositional detrital remanent magnetization* or pDRM. At some depth, magnetic grains become fixed. This depth varies from place to place and is controlled by (among other factors) the depth of bioturbation, clay content, and/or (magnetic) mineralogy. When sediments settle in the laboratory and undergo stirring to simulate the effect of bioturbation, the remanence has a strong linear dependence on the applied field, similar to TRM and CRM (although the remanence for a given field is lower by at least an order of magnitude). When sediments are deposited with no subsequant stirring, the remanence vectors of laboratory magnetizations are often shallower than the applied field, a phenomenon known as *inclination error*. The tangent of the observed inclination is usually some fraction ($\sim 0.4 - 0.6$) of the tangent of the applied field (King [1955], Tauxe and Kent [1984]). Thus, inclination error is at a maximum at inclinations near 45° and is negligible at high and low inclinations.

Interestingly, many natural sediments (e.g., deep sea or slowly deposited lake sediments) display no inclination error (e.g., Opdyke and Henry [1969]). In the laboratory, if sediments are disturbed (by tapping or stirring) after deposition, then the inclination error disappears (but the dependence of **M** on **H** remains). The tapping may be analogous to a pDRM in the deep sea which is acquired when the sediments are disturbed by bioturbation or rotation of particles in fluid-filled voids. Also, when sediments are allowed to settle "grain by grain" in the laboratory, they do not display inclination error (see e.g., Barton and McElhinny [1979]).

Tauxe et al. [1996a] compared the position of the Brunhes/Matuyama reversal boundary in marine carbonate cores with features in oxygen isotopic curves. They concluded that the magnetization was "locked-in" very close to the sediment-water interface at least in pelagic carbonates.

Furthermore, Hartl and Tauxe [1996] found that smoothing of the sedimentary record filters out only features with wavelengths less than a thousand years, while features longer than this are preserved. The slower the sedimentation rate, the greater the degree of smoothing. It should also be noted that, when sediments are squeezed in the laboratory to simulate compaction due to burial, the DRM becomes shallower (e.g., Anson and Kodama [1987]) and compaction related shallowing has been inferred in deep sea cores from depths greater than 100 m.

2.6.7. NATURAL REMANENT MAGNETIZATION

A rock collected from a geological formation has a magnetic remanence which may have been acquired by a variety of mechanisms some of which we have described. The remanence of this rock is called a *natural remanent magnetization* or NRM in order to avoid a genetic connotation in the absence of other compelling evidence. The NRM is often a combination of several components, each with its own history. The NRM must be examined in detail and the various remanence components must be carefully analyzed before their origin can be ascribed. The procedures for doing this are described later in the book.

2.7. Magnetic mineralogy

An essential part of every paleomagnetic study is a discussion of what is carrying the magnetic remanence and how the rocks got magnetized. For this, we need some knowledge of what the important natural magnetic

phases are, how to identify them, how they are formed, and what their
magnetic behavior is. We now provide a brief description of geologically
important magnetic phases. Useful magnetic characteristics of important
minerals can be found in Table 2.1.

Because iron is by far the most abundant transition element in the solar
system, most paleomagnetic studies depend on the magnetic iron species:
the iron-nickels (which are particularly important for extra-terrestrial magnetic studies), the iron-oxides such as magnetite, maghemite and hematite,
the iron-oxyhydroxides such as goethite, and the iron-sulfides such as greigite and pyrrhotite. We are concerned here with the latter three since iron-
nickel is very rare in terrestrial paleomagnetic studies.

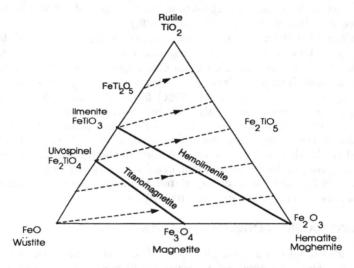

Figure 2.16. Ternary diagram for iron-oxides (redrawn from McElhinny [1973]). The
dashed lines with arrows indicate the direction of increasing oxidation (z). The solid lines
are solid solution series.

2.7.1. IRON-OXIDES

Two solid solution series are particularly important in paleomagnetism: the
ulvöspinel-magnetite and ilmenite-hematite series as shown on the ternary
diagram in Figure 2.16. Each of the solid lines in Figure 2.16 (labelled
Titanomagnetite and Hemoilmenite) represent increasing substitution of
titanium into the crystal lattices of magnetite and hematite respectively,
as one moves up and to the left along the lines. The amount of Ti substitution in titanomagnetites is denoted by "x", while substitution in the

hemoilmenites is denoted by "y". Values for x and y range from 0 (magnetite or hematite) to 1 (ulvöspinel or ilmenite). After crystallization, later oxidation will alter compositions along the dashed lines to the right and a titanomagnetite can become a titanomaghemite. The degree of oxidation is denoted by "z".

Titanomagnetites $Fe_{3-x}Ti_xO_4$

Magnetite (Fe_3O_4) has an inverse spinel structure (AB_2O_4). The oxygen atoms form a face-centered cubic lattice into which cations fit in either octahedral or tetrahedral symmetry. For each unit cell there are four tetrahedral sites (A) and eight octahedral sites (B). To maintain charge balance with the four oxygen ions (O^{2-}), there are two Fe^{3+} ions and one Fe^{2+} ion. Fe^{3+} has five unpaired spins, while Fe^{2+} has four. As discussed earlier, each unpaired spin contributes one Bohr magneton (m_b). The divalent iron ions all reside in the octahedral lattice sites, whereas the trivalent iron ions are split evenly between octahedral and tetrahedral sites: $Fe^{3+}|Fe^{3+}Fe^{2+}|O_4$. The A and B lattice sites are coupled with antiparallel spins and magnetite is ferrimagnetic. Therefore, the net moment of magnetite is 4 m_b per molecule (at 0 K).

Titanomagnetites occur as primary minerals in igneous rocks. Magnetite, as well as various members of the hemoilmenite series, can also form as a result of high temperature oxidation. In sediments, magnetite often occurs as a detrital component, but it can also be produced by bacteria or authigenically during diagenesis.

Substitution of Ti^{4+}, which has no unpaired spins, has a profound effect on the magnetic properties of the resulting titanomagnetite. Ti^{4+} substitutes for a trivalent iron ion. In order to maintain charge balance, another trivalent iron turns into a divalent iron ion. The end members of the solid solution series are:

magnetite	ulvöspinel
$Fe^{3+}|Fe^{3+}Fe^{2+}|O_4$	$Fe^{2+}|Fe^{2+}Ti^{4+}|O_4$
$x = 0$	$x = 1$

Ulvöspinel is antiferromagnetic because the A and B lattice sites have the same net moment. When x is between 0 and 1, the mineral is called a titanomagnetite. If x is 0.6, the mineral is called TM60. Above 600°C, there is complete solid solution between ulvöspinel and magnetite. Below this temperature, the spinel can exsolve into two phases, one with a high titanium content and the other with a low titanium content.

Titanium substitution has the effect of: 1) increasing the cell dimensions, 2) decreasing the Curie temperature, 3) decreasing saturation magnetization and susceptibility, and 4) slightly increasing coercivity. The diagnostic properties of magnetite are listed in Table 2.1.

The large M_s of magnetite (see Table 2.1) means that for deviations from equant grains as small as 10%, the magnetic anisotropy energy becomes dominated by shape. Nonetheless, aspects of the magnetocrystalline anisotropy provide useful diagnostic tests. The magnetocrystalline anisotropy constants are a strong function of temperature. On warming to $\sim$-100°C from near absolute zero, changes in these constants can lead to an abrupt loss of magnetization, which is known loosely as the *Verwey transition*. Identification of the Verwey transition suggests a remanence that is dominated by magnetocrystalline anisotropy. Furthermore the temperature at which it occurs is sensitive to oxidation and the transition can be completely supressed by maghemitization (see Özdemir et al. [1993]).

Hematite-Ilmenite $Fe_{2-y}Ti_yO_3$

Hematite has a corundum structure. It is rhombohedral with a pseudocleavage (perpendicular to the c axis); hematite tends to break into flakes. It is antiferromagnetic, with a weak parasitic ferromagnetism resulting from either spin-canting or defect ferromagnetism (see Figure 2.4). Above about -10°C (the *Morin transition*), the magnetization is constrained by aspects of the crystal structure to lie perpendicular to the c axis. Below the Morin transition, spin-canting disappears and the magnetization is parallel to the c axis. This effect could be used to demagnetize the grains dominated by spin-canting but it does not affect the defect moments.

Hematite occurs widely in oxidized sediments and dominates the magnetic properties of red beds. It occurs as a high temperature oxidation product in certain igneous rocks. Depending on grain size, among other things, it is either black (specularite) or red (pigmentary). Diagnostic properties of hematite are listed in Table 2.1.

Oxidation of (titano)magnetites to (titano)maghemites

The titanomagnetite series of minerals is subject to low temperature oxidation. The resulting mineral has an "open" spinel structure and is called titanomaghemite. Maghemite has the same composition as hematite but it has a spinel structure, with vacancies left from the diffusion of iron ions out of the crystal.

Maghemite is metastable and inverts with time and temperature to the more compact hematite structure. Inversion of natural maghemite is usually complete by about 350°C, but it can survive until much higher temperatures (see e.g., Özdemir and Banerjee [1984]). Maghemitization results in a decrease in saturation magnetization and an increase in Curie temperature.

2.7.2. IRON-OXYHYDROXIDES AND IRON-SULFIDES

Of the many iron oxyhydroxides that occur in any abundance in nature, only goethite (αFeOOH) is magnetic. It is antiferromagnetic with what is most likely a defect magnetization. It occurs widely as a weathering product of iron-bearing minerals and as a direct precipitate from iron-bearing solutions. It is metastable under many conditions and dehydrates to hematite with time or elevated temperature. Dehydration is usually complete by about 325°C. It is characterized by a very high coercivity but a low Néel temperature of about 100–150°C. Diagnostic properties of goethite are listed in Table 2.1.

There are two iron-sulfides that are important to paleomagnetism: greigite (Fe_3S_4) and pyrrhotite (Fe_7S_8-$Fe_{11}S_{12}$). These are ferrimagnetic and occur in reducing environments. They both tend to oxidize to various iron oxides leaving paramagnetic pyrite as the sulfide component. The Curie temperature of pyrrhotite is about 325°C (see Table 2.1) and the maximum unblocking temperature of greigite is approximately 330°C. Other diagnostic properties of greigite and pyrrhotite are listed in Table 2.1.

2.8. Tools to constrain magnetic mineralogy

We now have an idea of how rocks might become and stay magnetized and what minerals might carry the magnetic remanence. Different minerals imply different modes of remanence acquisition, which have in turn profound implications for the probable reliability of the magnetic record. Therefore, it is useful to be able to identify the magnetic minerals in the rock.

There are many tools for mineral identification: optical and electron microscopy, x-ray diffraction, electron probe microanalysis, etc. Here we consider selected rock magnetic methods.

2.8.1. CURIE TEMPERATURE ESTIMATION

As is clear from a glance at Table 2.1, the Curie temperature can be diagnostic of magnetic mineralogy. Curie temperatures are measured by applying a large field and measuring the response of a sample as a function of temperature. Ferro- and paramagnetic substances will be drawn into a magnetic field, while diamagnetic substances are repulsed. A Curie balance compensates for the attractive (repulsive) force of the field gradient on the sample by generating a field which nulls the attraction exactly. The current required to do this is proportional to the magnetization of the sample. Two so-called *thermomagnetic curves* generated in this manner are shown in Figure 2.17.

[See Example 2.1]

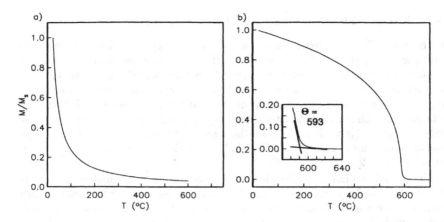

Figure 2.17.   Magnetization as a function of temperature for: a) a paramagnetic substance, and b) a ferromagnetic (*sensu lato*) substance. The "intersecting tangents" method for estimating Curie temperatures is shown in the inset.

As might be expected from our theoretical treatment of paramagnetism, paramagnetic substances have magnetizations that are proportional to T^{-1} (see Figure 2.17a). Ferromagnetic samples produce behavior something like that shown in Figure 2.17b, as would be expected from equation 2.12.

Estimating the Curie temperature is not as simple as it seems at first glance. Grommé et al. [1969] proposed the use of the intersection point of the two tangents to the thermomagnetic curve that bounds the Curie temperature, as shown in Figure 2.17b. The *intersecting tangents method* is straightforward to do by hand, but is rather subjective and is difficult to automate. Moskowitz [1981] developed a method for extrapolating the ferromagnetic behavior expected from experimental data through the Curie temperature to determine the point at which the ferromagnetic contribution reaches zero.

A third method for estimating Curie temperatures from thermomagnetic data, the *differential method*, seeks the maximum curvature in the thermomagnetic curve. This method is shown in Figure 2.18. First, we calculate the derivative (dM/dT) of the data in Figure 2.17b (see Figure 2.18a). Then, these data are differentiated once again to produce d^2M/dT^2 (Figure 2.18b). The maximum in the second derivative occurs at the point of maximum curvature in the thermomagnetic curve and is a reasonable estimate of the Curie temperature.

The principal drawback of the differential method of Curie temperature estimation is that noise in the data is greatly amplified by differentiation,

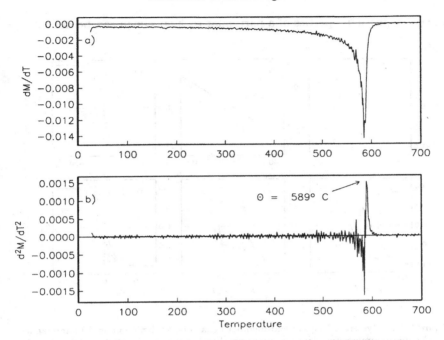

Figure 2.18. Data from Figure 2.17b differentiated: a) once and b) twice. The Curie temperature (Θ) is calculated as the maximum in the second derivative curve. Temperature is in °C.

which makes identification of the Curie temperature difficult. The drawbacks of the differential method can often be overcome by smoothing the data either by calulating three or more point running means, or using some filter either by Fourier methods or in the temperature domain. We find that a triangular filter works well.

[See Example 2.2]

2.8.2. INTERPRETATION OF HYSTERESIS LOOPS

Geological materials can be magnetically complicated. The magnetization of a natural crystal may not be uniform, particularly if it is in the PSD or MD state. The moments may be strongly affected by thermal energy and they may display superparamagnetic (SP) behavior, or they may be mixtures of a wide variety of naturally occurring magnetic phases with very different behavior. Furthermore, there may also be diamagnetic or paramagnetic material in the sample. As a result, determining the controls on the magnetic properties of a particular rock specimen can be difficult.

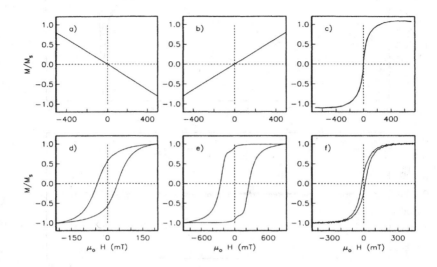

Figure 2.19. Hysteresis loops of end-member behaviors: a) diamagnetic, b) paramagnetic, c) superparamagnetic (data for submarine basaltic glass), d) uniaxial, single domain (data from Tauxe et al. [1996b]), e) magnetocrystalline, single domain (data from Tauxe et al. [1996b]), f) pseudo-single domain (data for marine carbonate).

Because hysteresis behavior is strongly controlled by mineralogy and grain size, hysteresis loops have the potential to help constrain the makeup of a given rock specimen. The hysteresis loop of a given sample will be the sum of all the curves generated by the individual grains. Each population of grains with a consistent coercivity spectrum will leave its imprint on the resulting loop. We illustrate some of the building blocks of possible hysteresis loops in Figure 2.19. Figure 2.19a shows the negative slope typical of diamagnetic material such as carbonate or quartz, while Figure 2.19b shows a paramagnetic slope. Such slopes are common when the sample has little ferromagnetic material and is rich in iron-bearing phases such as biotite or clay minerals.

When grain sizes are very small, a sample can display superparamagnetic "hysteresis" behavior (Figure 2.19c). The SP curve follows a Langevin function $L(a)$ (see equation 2.8) where a is $M_s v \mu_o H / kT$.

Above some critical volume, grains will have relaxation times that are sufficient to retain a stable remanence. As discussed earlier in the chapter, populations of randomly oriented stable grains can produce hysteresis loops with a variety of shapes, depending on the origin of magnetic anisotropy and domain state. We show loops from samples that illustrate several typical end-member styles of hysteresis behavior in Figure 2.19d-f.

Figure 2.19d shows a loop characteristic of samples whose remanence stems from SD magnetite with uniaxial anisotropy. In Figure 2.19e, we show data from specular hematite whose anisotropy is magnetocrystalline in origin (hexagonal within the basal plane). Note the very high M_r/M_s ratio of nearly one. Finally, we show a loop that is generally interpreted to indicate the dominance of PSD (Figure 2.19f).

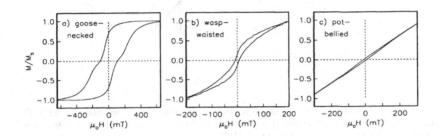

Figure 2.20. Hysteresis behavior of various mixtures: a) magnetite, and hematite, b) SD/SP magnetite (data from Tauxe et al. [1996b]), c) another example of SD/SP magnetite.

In the messy reality of geological materials, we often encounter mixtures of several magnetic phases and/or domain states. Such mixtures can lead to distorted loops, such as those shown in Figure 2.20. In Figure 2.20a, we show a mixture of hematite plus SD-magnetite. The loop is distorted in a manner that we refer to as *goose-necked*. Another commonly observed mixture is SD plus SP magnetite (see Tauxe et al. [1996b]) which results in loops that are either *wasp-waisted* (see Figure 2.20b) or *pot-bellied* (see Figure 2.20c). Another effect of mixing SP with SD grains is a suppression of the M_r/M_s ratio, so the ratios M_r/M_s and H_{cr}/H_c plot along a different slope in the "Day plot" than those expected for particles that range from SD to MD size. (see Figure 2.15). In Figure 2.21, we compare the effect of increasing the population of SP grains on the hysteresis ratios (from the simulations of

Tauxe et al. [1996b]) with that of increasing grain size (data of Day et al. [1977]).

Considering the loops shown in Figure 2.20, we immediately notice that there are two distinct causes of loop distortion: mixing two phases with different coercivities (see Wasilewski [1973], Roberts et al. [1995]), and mixing

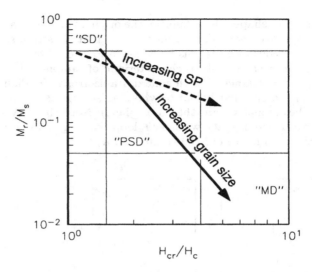

Figure 2.21. Comparison of trends from simulated loops for populations of SD/SP magnetite (from Tauxe et al. [1996b]) with those measured for populations of magnetite grains of known grain size (from Day et al. [1977]).

SD and SP domain states (see Pick and Tauxe [1994]). We differentiate the two types of distortion as "goose-necked" and "wasp-waisted" (see Figure 2.20) because they look different and they mean different things. Because the ΔM and $d\Delta M/dH$ curves are sensitive only to the remanence carrying phases, and not, for example, to the SP fraction, we can use these curves to distinguish the two sources of distortion. In Figure 2.22, we show several representative loops, along with the ΔM and $d\Delta M/dH$ curves. Distortion resulting from two phases with different coercivities (e.g., hematite plus magnetite or two distinct grain sizes of the same mineral) results in a "two humped" $d\Delta M/dH$ curve, whereas wasp-waisting which results from mixtures of SD + SP populations have only one "hump".

2.8.3. FOURIER ANALYSIS OF HYSTERESIS LOOPS

In practice, hysteresis measurements may yield rather noisy data. Jackson et al. [1990] suggested that noisy hysteresis data could be filtered using a Fourier transform. The advantages of Fourier smoothing are that the calculated hysteresis parameters are less sensitive to noise and that the ΔM and $d\Delta M/dH$ curves are more readily interpreted.

The steps involved in Fourier smoothing of hysteresis loops are as follows (see Figure 2.23):

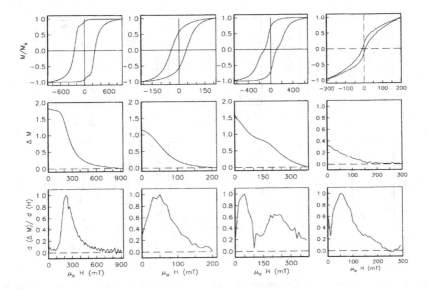

Figure 2.22. Top panels: hysteresis curves, middle panels: ΔM curves and bottom panels: $d\Delta M/dH$ curves. From the left to right: hematite, SD magnetite, hematite plus magnetite, and SD plus SP magnetite.

[See Example 2.3]

• First, the contribution of paramagnetic and diamagnetic phases must be removed. Figure 2.23a shows some typical data from carbonate rich sediments. These samples have a strong diamagnetic (negative high field slope) contribution. We remove the diamagnetic contribution by calculating a best-fit line using linear regression for the data at high fields (after the ferromagnetic phases have reached saturation) and removing its contribution by subtraction (see Figure 2.23b).

• In order to ensure uniformity of data treatment, Jackson et al. [1990] recommend truncating the data at some fixed percentage of M_s (after slope adjustment). We truncate the data at 99.9% of M_s in Figure 2.23b.

• A Fourier transform requires data with a single y value for every x value and hysteresis data, as normally plotted are not suitable. The loops can be mapped into a suitable form for Fourier analysis by transforming the field values into radians, as shown in Figure 2.23c. The unfolded loop starts at the point when the descending curve intersects

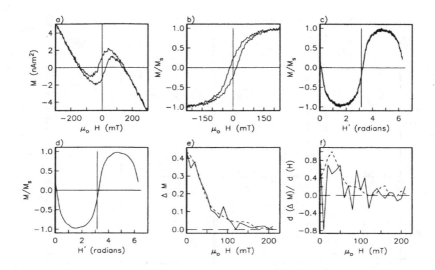

Figure 2.23. Steps in Fourier smoothing: a) the original data - note the negative high field slope from the diamagnetic contribution of carbonate, b) data with the high field slope removed and truncated to 99.9% of the maximum value of M, c) data from b) "unfolded" into radians, d) data from c) smoothed by using the first 15 terms of Fourier series, e) comparison of ΔM curve using original data (solid line) and smoothed data (dashed line), f) comparison of the $d\Delta M/dH$ curve using the original data (solid line) and smoothed data (dashed line).

the y axis (M_r). From $H = 0 \rightarrow -H_{max}$, H is mapped linearly to radians ($H' = 0 \rightarrow \pi/2$). From $H = -H_{max} \rightarrow 0$, H is mapped to $H' = \pi/2 \rightarrow \pi$. From $H = 0 \rightarrow +H_{max}$, we map H to $H' = \pi \rightarrow 3\pi/2$, and finally, for $H = +H_{max} \rightarrow 0$, H is converted to $H' = 3\pi/2 \rightarrow 2\pi$.

• The "unfolded" data can then be subjected to a Fourier Transform as described by Jackson et al. [1990]. The data can be smoothed by retaining only a specified number of terms (see Figure 2.23d). Finally, hysteresis parameters can be calculated from the reconstituted loop and ΔM and $d\Delta M/dH$ curves can be plotted (see Figure 2.23e-f).

2.9. Summary

This chapter has dealt with the nitty gritty of rock magnetism. We have explored how rocks get and stay magnetized. We have also discussed rock magnetic methods for identifying magnetic minerals. In the following chapter we turn our attention to sampling techniques and we discuss what to do with samples once we have them in the laboratory.

TABLE 2.1. Physical properties of magnetic minerals.

Magnetite	Fe_3O_4
Density = 5197 kg m^{-3}	Dunlop and Özdemir [1997]
Curie temperature = 580°C	Dunlop and Özdemir [1997]
Saturation Magnetization = 92 Am^{2}kg^{-1}	O'Reilly [1984]
Anisotropy Constant = -2.6 Jkg^{-1}	Dunlop and Özdemir [1997]
Volume susceptibility = $\sim$ 1 SI	O'Reilly [1984]
Typical coercivities are 10's of mT	O'Reilly [1984]
Verwey transition: 110-120 K	Özdemir and Dunlop [1993]
Cell edge = 83.96 nm	Dunlop and Özdemir [1997]
Maghemite	γFe_2O_3
Density = 5074 kg m^{-3}	Dunlop and Özdemir [1997]
Curie temperature = 590-675°C	Dunlop and Özdemir [1997]
Saturation Magnetization = 74 Am^{2}kg^{-1}	Dunlop and Özdemir [1997]
Anisotropy Constant = 0.92 Jkg^{-1}	Dunlop and Özdemir [1997]
Verwey transition: suppressed	Dunlop and Özdemir [1997]
Breaks down to αFe_2O_3: between 250$\rightarrow$ 750°C	Dunlop and Özdemir [1997]
TM60	$Fe_{2.4}Ti_{0.6}O_4$
Density = 4939 kg m^{-3}	Dunlop and Özdemir [1997]
Curie temperature = 150°C	Dunlop and Özdemir [1997]
Saturation Magnetization = 24 Am^{2}kg^{-1}	Dunlop and Özdemir [1997]
Anisotropy Constant = 0.41 Jkg^{-1}	Dunlop and Özdemir [1997]
Coercivity $\sim$ 8 mT	Dunlop and Özdemir [1997]
Verwey transition: suppressed	Dunlop and Özdemir [1997]
Cell edge = 84.82 nm	Dunlop and Özdemir [1997]
Hematite	αFe_2O_3
Density = 5271 kg m^{-3}	Dunlop and Özdemir [1997]
Néel temperature = 675°C	O'Reilly [1984]
Saturation Magnetization = 0.4 Am^{2}kg^{-1}	O'Reilly [1984]
Anisotropy Constant = 228 Jkg^{-1}	Dunlop and Özdemir [1997]
Volume susceptibility = $\sim$ 1.3 x 10^{-3} SI	O'Reilly [1984]
Coercivities vary widely and can be 10's of teslas	Banerjee [1971]
Morin Transition: $\sim$ 250-260 K (for > 0.2 μm)	O'Reilly [1984]

TABLE 2.1 - continued

Goethite	αFeOOH
Density = 4264 kg m^{-3}	Dunlop and Özdemir [1997]
Néel temperature: 70 $\to$ 125°C	O'Reilly [1984]
Saturation Magnetization = 10^{-3} $\to$ 1 Am^{2}kg^{-1}	O'Reilly [1984]
Anisotropy Constant = 0.25 $\to$ 2 Jkg^{-1}	Dekkers [1989]
Volume susceptibility = $\sim$ 1 x 10^{-3} SI	Dekkers [1989]
Coercivities can be 10's of teslas	
Breaks down to hematite: 250 $\to$ 400°C	

Pyrrhotite	Fe$_7$S$_8$
Density = 4662 kg m^{-3}	Dunlop and Özdemir [1997]
Curie temperature = $\sim$ 325°C	Dekkers [1989a]
Saturation Magnetization = 0.4 -$\sim$ 20 Am^{2}kg^{-1}	Worm et al. [1993]
Volume susceptibility = $\sim$ 1 x 10^{-3} $\to$ 1 SI	Collinson [1983];O'Reilly [1984]
Anisotropy Constant = 20 Jkg^{-1}	O'Reilly [1984]
Coercivities vary widely and can be 100's of mT	O'Reilly [1984]
Has a transition at $\sim$ 34 K	Dekkers et al. [1989]
	Rochette et al. [1990]
Breaks down to magnetite: $\sim$ 500°C	Dunlop and Özdemir [1997]

Greigite	Fe$_3$S$_4$
Density = 4079 kg m^{-3}	Dunlop and Özdemir [1997]
Maximum unblocking temperature = $\sim$ 330°C	Roberts [1995]
Saturation Magnetization = $\sim$ 25 Am^{2}kg^{-1}	Spender et al. [1972]
Anisotropy Constant = -0.25 Jkg^{-1}	Dunlop and Özdemir [1997]
Coercivity 60$\to$> 100 mT	Roberts [1995]
Has high M_r/χ ratios $\sim$ 70 x 10^3 Am^{-1}	Snowball and Thompson [1990]
Breaks down to magnetite: $\sim$ 270-350°C	Roberts [1995]

2.10. Examples

• Example 2.1

Plot the thermomagnetic data in file **ex2.1** using **plotxy**. Then convert to absolute temperature and find the inverse. Plot these data using **plotxy**.
Solution:
Type the boldfaced commands (the computer responses are in normal typeface):

% plotxy
Enter commands for graph 1
file ex2.1
read
xlabel Temperature
ylabel M
plot 1 1

PostScript file written to: mypost
========== ——————————————
Enter commands for graph 2
stop

These commands cause **plotxy** to generate the postscript file **mypost** as shown in Figure 2.24.
Now type:
% nawk '{print 1/($1+273),$2}' ex2.1 > ex2.1a
The Unix utility **nawk** is similar to **awk** and performs a number of spreadsheet like operations on data files.
To plot the inverted temperatures, type:

% plotxy
Enter commands for graph 1
file ex2.1a
read
xlabel Inverse Temperature
ylabel M
plot 1 1

PostScript file written to: mypost
========== ——————————————
Enter commands for graph 2
stop

This generates the postcript file **mypost** as shown in Figure 2.25.

• Example 2.2

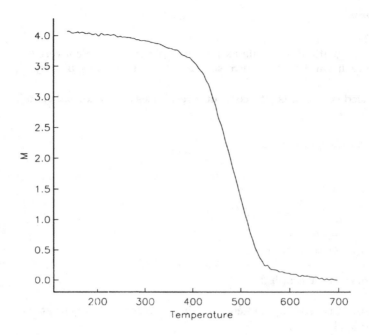

Figure 2.24. Thermomagnetic curve for magnetite in file **ex2.1**, plotted using **plotxy**. See Example 2.1.

Use the program **curie** to calculate the Curie temperature of the data contained in file **ex2.1**. First scan through a range of smoothing intervals from 1 to 100° , and then choose the optimal smoothing interval (the smallest interval necessary to isolate the correct peak in the second derivative). Finally, repeat this, but truncate the data set to between 400° to 600°.
Solution:
First type:
% **curie -s 3 100 1 < ex2.1**
which produces a list of Curie temperatures for smoothing intervals from 3 to 100° at 1° invervals:

    Tc = 203
       3 203
       4 203
       5 203
       6 549
       7 549

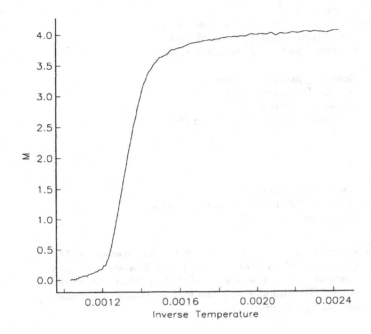

Figure 2.25. Data from file **ex2.1** plotted as inverse absolute temperature using **plotxy** (see Example 2.1.)

⋮

Now type:

% **curie -lsp 20 3 100 1 < ex2.1 | plotxy**

This causes **plotxy** to produce a postscript file **mypost** shown in Figure 2.26, and the output Curie temperature is 538°C.

It may be that there are several Curie temperatures in a particular data set and that you wish to focus in on one of them. The **curie** program allows truncation of the data set to a given interval by specifying **Tmin** and **Tmax** with the -**t** option. As usual, check Appendix 1, or type **curie -h** :

% **curie -h**

Usage: curie -[lspt] [smooth] [low hi step] [Tmin Tmax] Standard I/O

      -l low pass filter using smoothing interval [smooth]

        NB: [smooth] must be an odd number ≥ 3

      -s scan range of frequencies

        [low] to [hi] using a spacing of step

[low], [hi] and [step] must be odd
-p plot option on to generate Plotxy command file
    can be piped directly to plotxy and viewed:
    curie -p < filename | plotxy; ghostview mypost
      printed:
    curie -p < filename | plotxy; lpr mypost
      or saved to a file for modification:
    curie -p < filename > eqarea.com
-t truncates to interval between [Tmin] and [Tmax]
defaults are:
    no smoothing
    plot option off
    uses entire record

To complete the exercise, type the following:

% **curie -lspt 20 3 100 1 400 600 < ex2.1 | plotxy**

View the **mypost** file, as in Figure 2.27.

• Example 2.3

Use the program **hystcrunch** to calculate the hysteresis parameters, M_r, M_s, H_c, H'_{cr} and H''_{cr} from the data in file **ex2.3**. These data are in the raw format that is the output from the available Micromag 2900 alternating gradient magnetometer. NB: retaining only the first 30 Fourier harmonics results in a smoother $d\Delta M/dH$ curve.

Solution

Type the following:

% **hystcrunch -mt 30 < ex2.3 | plotxy**

This causes plotxy to produce a postcript file (**mypost**, see Figure 2.28) which can be viewed or plotted as desired.

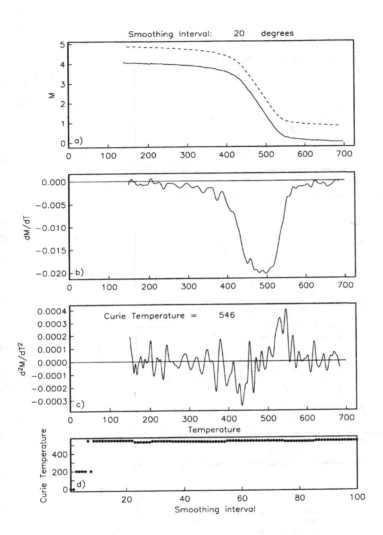

Figure 2.26. Plot generated in Example 2.2 by the command: **curie -lsp 20 3 100 1 < ex2.1 | plotxy.** a) The data in file **ex2.1** plotted as solid line and smoothed data using a smoothing interval of 20° shown offset upwards as a dashed line; b) the first derivative of the smoothed data; c) the second derivative of the smoothed data; d) the curie temperature as a function of smoothing interval. Note that the solution doesn't stabilize until 20° and the interval 6-19 results in an overestimate of the Curie temperature.

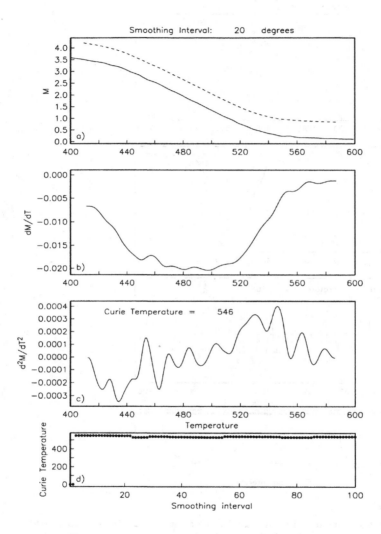

Figure 2.27. Plot generated in Example 2.2 using the command: **curie -lspt 20 3 100 1 400 600 < ex2.1 | plotxy**. The data are the same as in Example 2.1, but are truncated to between 400 and 600° C.

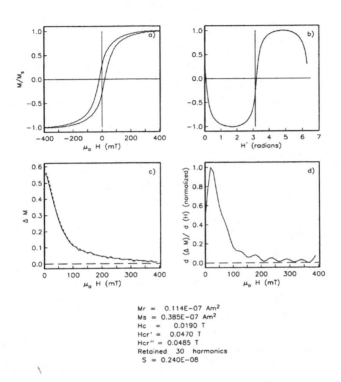

Figure 2.28. Output of command: **hystcrunch -mt 30 < ex2.3 | plotxy** in Example 2.3. a) The data with slope correction (S) subtracted, b) data from a) unfolded into radians, c) ΔM curves of data from a) (solid line) and Fourier smoothed data (dashed line), and d) the derivative of the dashed line in c).

Chapter 3

PALEOMAGNETIC PROCEDURES

As discussed in the previous chapter, rocks become magnetized in a variety of ways. Both igneous and sedimentary rocks can be affected by chemical change, thereby acquiring a secondary magnetization. Many magnetic materials are affected by viscous remanent magnetization. The various components of magnetization sum together to constitute the NRM which is the "raw" remanence of the sample after extraction. The goal of paleomagnetic laboratory work is to isolate the various components of remanence and to ascribe origin, age and reliability to the components. Before the laboratory work can begin, however, samples must be collected. Sampling strategy is crucial to a successful study. We will briefly describe techniques for sampling, methods of orientation and overall philosophy. We will then turn to an overview of some of the more useful field and laboratory techniques. The reader is referred to Butler [1992] for an excellent introductory book on paleomagnetism.

3.1. Paleomagnetic sampling

There are several goals in sampling rock units. One is to average the errors involved in the sampling process itself. Another is to assess the reliability of the geological recording medium. In addition, we often wish to average the scatter caused by secular variation of the geomagnetic field in order to estimate the time-averaged paleomagnetic field direction representative of the time that the rock unit acquired its magnetization.

The objectives of averaging geological and sampling "noise" are achieved by taking a number N of individually oriented *paleomagnetic samples* from a single rock unit (called a *paleomagnetic site*). The most careful sample orientation procedure has an uncertainty of some 3°. Precision is gained proportional to $\sqrt{N}$, so to improve the precision from 3° to 1°, nine individually oriented samples are required. The number of samples taken should be tailored to the particular project at hand. If one wishes to know polarity, perhaps three samples would be sufficient (these would be taken primarily to assess "geological noise"). If, on the other hand, one wished to make inferences about secular variation of the geomagnetic field, more samples would be necessary to suppress sampling noise.

Some applications in paleomagnetism require that the secular variation of the geomagnetic field (the paleomagnetic "noise") be averaged in order to determine the time-averaged field direction. The geomagnetic field varies with time constants ranging from milliseconds to millions of years. It is a reasonable first order approximation to assume that, when averaged over, say, 10,000 years, the geomagnetic field is that of a geocentric axial dipole (equivalent to the field that would be produced by a bar magnet at the center of the Earth, aligned with the spin axis; see Chapter 1). Thus, when

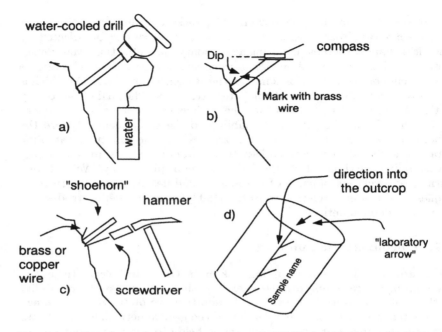

Figure 3.1. Sampling technique with a water-cooled drill: a) drill the sample, b) insert a non-magnetic slotted tube with an adjustable platform around the sample. Rotate the slot to the top of the sample and note the azimuth and plunge of the drill direction (into the outcrop) with a sun and/or magnetic compass and inclinometer. c) Mark the sample through the slot with a brass or copper wire and extract the sample. d) Make a permanent arrow on the side of the sample in the direction of drill and label the sample with the sample name. Make a note of the name and orientation of the arrow in a field notebook. Finally, slice the sample into a standard length (e.g., one inch). Measurements in the author's laboratory are made with respect to a "laboratory arrow" the orientation of which can be calculated from the notebook values of the field arrow drawn in the direction of drill into the outcrop, knowing the relationship of the field and laboratory arrows.

a time-averaged field direction is required, enough sites must be sampled to span sufficient time to achieve this goal. A good rule of thumb is about ten sites (each with nine to ten samples), spanning 10,000 years.

Samples can be taken in the field using a gasoline or electric powered drill or as "hand samples". The samples must be oriented before they are removed from the rock unit. There are many ways to orient samples and possible conventions are shown in Figures 3.1 and 3.2.

If a magnetic compass is used to orient samples in the field, the measured azimuth must be adjusted by the local magnetic declination, which

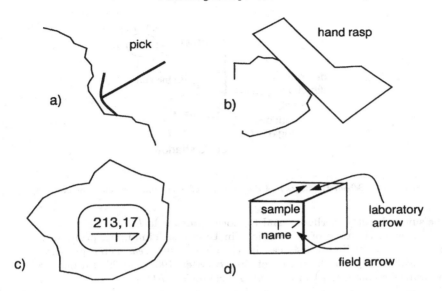

Figure 3.2. Hand sampling technique: a) dig down to fresh material, b) rasp off a flat surface, c) mark the strike and dip on the sample, d) grind the sample into a standard cubic sample (e.g., a nominal cubic inch), and label it with a "laboratory arrow" on it and the sample name.

can be calculated from the known reference field (IGRF or DGRF; see Example 1.3 and Chapter 1). Calculation of a direction using a sun compass is more involved. A dial with a vertical needle (a "gnomon") is placed on the horizontal platform shown in Figure 3.3. The angle (α) that the sun's shadow makes with the drilling direction is noted as well as the exact time of sampling and the location of the sampling site. With this information and the aid of the Astronomical Almanac or a simple algorithm, it is possible to calculate the desired direction to reasonable accuracy (the biggest cause of uncertainty is actually reading the shadow angle!).

[See Example 3.1]

Referring to Figure 3.3, we see that the azimuth of the desired direction is the direction of the of the shadow plus the shadow angle α. The declination of the shadow itself is 180° from the direction toward the sun. In Figure 3.4, the problem of calculating declination from sun compass information is set up as a spherical trigonometry problem, similar to those introduced in Chapter 1. The declination of the shadow direction β', is given by 180 - β. We also know the latitude of the sampling location L (λ_L). We need to calculate the latitude of S (the point on the Earth's surface where

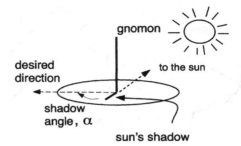

Figure 3.3.   Schematic diagram of the principle of operation of a sun compass.

the sun is directly overhead), and the local hour angle H.

Knowing the time of observation (in Universal Time), the position of S ($\lambda_s = \delta, \phi_s$ in Figure 3.4) can be calculated with reasonable precision (to within 0.01°) for the period of time between 1950 and 2050 using the procedure recommended in the 1996 Astronomical Almanac:

• First, calculate the Julian Day J. Then, calculate the fraction of the day in Universal Time U. Finally, calculate the parameter d which is the number of days from J2000 by:

$$d = J - 2451545 + U.$$

• The mean longitude of the sun (ϕ_s), corrected for aberration, can be estimated in degrees by:

$$\phi_s = 280.461 + 0.9856474d.$$

• The mean anomaly $g = 357.528 + 0.9856003d$ (in degrees).
• Put ϕ_s and g in the range $0 \rightarrow 360°$.
• The longitude of the ecliptic is given by $\phi_E = \phi_s + 1.915 \sin g + 0.020 \sin 2g$ (in degrees).
• The obliquity of the ecliptic is given by $\epsilon = 23.439 - 0.0000004d$.
• Calculate the right ascension (A) by:

$$A = \phi_E - ft \sin 2\phi_E + (f/2)t^2 \sin 4\phi_E,$$

where $f = 180/\pi$ and $t = \tan^2 \epsilon/2$.
• The so-called "declination" of the sun (δ in Figure 3.4 which should not be confused with the magnetic declination D), which we will use as the latitude λ_s, is given by:

$$\delta = \sin^{-1}(\sin \epsilon \sin \phi_e).$$

- Finally, the equation of time in degrees is given by $E = 4(\phi_s - A)$.

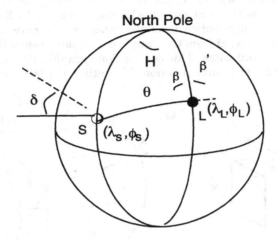

Figure 3.4. Calculation of the azimuth of the shadow direction (β') relative to true North, using a sun compass. L is the site location (at λ_L, ϕ_L), S is the position on the Earth where the sun is directly overhead (λ_S, ϕ_S).

We can now calculate the Greenwich Hour Angle GHA from the Universal Time U (in minutes) by $GHA = (U + E)/4 + 180$. The local hour angle (H in Figure 3.4) is $GHA + \phi_L$. We calculate β using the laws of spherical trigonometry (see Chapter 1). First we calculate θ by the Law of Cosines (remembering that the cosine of the colatitude equals the sine of the latitude):

$$\cos \theta = \sin \lambda_L \sin \lambda_s + \cos \lambda_L \cos, \lambda_s \cos H$$

and finally using the Law of Sines:

$$\sin \beta = (\cos \lambda_s \sin H)/\sin \theta.$$

If $\lambda_s < \lambda_L$, then the required angle is the shadow direction β', given by: $\beta' = 180 - \beta$. The azimuth of the desired direction in Figure 3.3 is β' plus the measured shadow angle α.

3.2. Transformation of coordinate systems

Samples are brought to the laboratory and trimmed into standard sizes and shapes. These sub-samples are called *paleomagnetic specimens*. The sample coordinate system is defined by a right-hand rule where the thumb ($\mathbf{X}_1$) is directed parallel to an arrow marked on the sample, the index finger ($\mathbf{X}_2$) is in the same plane but at right angles and clockwise to $\mathbf{X}_1$ and the middle finger ($\mathbf{X}_3$) is perpendicular to the other two (Figure 3.5a). Data often must be transformed from the sample coordinate system into, for example, geographic coordinates. This can be done graphically with a stereonet or by means of matrix manipulation. We outline the latter method here.

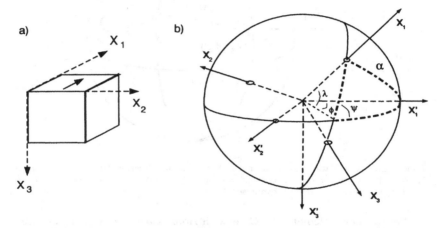

Figure 3.5. a) Sample coordinate system. b) Trigonometric relations between two cartesian coordinate systems, $\mathbf{X}_i$ and $\mathbf{X}'_i$. λ, ϕ, ψ are all known and the angles between the various axes can be calculated using spherical trigonometry. For example, the angle α between $\mathbf{X}_1$ and $\mathbf{X}'_1$ forms one side of the triangle shown by dash-dot lines. Thus, $\cos \alpha = \cos \lambda \cos \phi + \sin \lambda \sin \phi \cos \psi$.

The transformation of coordinates (x_i) from the $\mathbf{X}_i$ axes to the coordinates in the desired $\mathbf{X}'$ coordinate system (x'_i) is done by $x'_i = a_{ij}x_j$, or:

$$\begin{pmatrix} x'_1 \\ x'_2 \\ x'_3 \end{pmatrix} = \begin{pmatrix} a_{11} & a_{12} & a_{13} \\ a_{21} & a_{22} & a_{23} \\ a_{31} & a_{32} & a_{33} \end{pmatrix} \begin{pmatrix} x_1 \\ x_2 \\ x_3 \end{pmatrix}, \tag{3.1}$$

where the a_{ij} are the *direction cosines* (the cosines of the angles between the different axes), where the subscript i refers to the new coordinate system $\mathbf{X}'$ and the subscript j refers to the old $\mathbf{X}$ coordinates. Thus, a_{12} is the cosine of the angle between $\mathbf{X}_1'$ and $\mathbf{X}_2$. The various a_{ij} can be calculated using spherical trigonometry (Chapter 1). For example, a_{11} for the general case depicted in Figure 3.5 is $\cos\alpha$, which is given by the Law of Cosines (see Chapter 1) by using appropriate values, or:

$$\cos\alpha = \cos\lambda\cos\phi + \sin\lambda\sin\phi\cos\psi.$$

The other a_{ij} can be calculated in a similar manner. In the case of most coordinate system rotations used in paleomagnetism, X_2 is in the same plane as X_1' and X_2' (and is horizontal) so $\psi = 90°$. This problem is much simpler. The directions cosines for the case where $\psi = 90$ are:

$$a = \begin{pmatrix} \cos\lambda\cos\phi & -\sin\phi & -\sin\lambda\cos\phi \\ \cos\lambda\sin\phi & \cos\phi & -\sin\lambda\sin\phi \\ \sin\lambda & 0 & \cos\lambda \end{pmatrix}. \tag{3.2}$$

The new coordinates can be obtained from equation 3.1, as follows:

$$\begin{aligned} x_1' &= a_{11}x_1 + a_{12}x_2 + a_{13}x_3 \\ x_2' &= a_{21}x_1 + a_{22}x_2 + a_{23}x_3 \\ x_3' &= a_{31}x_1 + a_{32}x_2 + a_{33}x_3. \end{aligned} \tag{3.3}$$

The declination and inclination can be calculated by inserting these values in equation 1.3 in Chapter 1.

In practice, there are two transformations that are routinely made in paleomagnetism. Magnetizations are measured in sample coordinates. First, they must then be rotated into geographic coordinates. For this, the azimuth and plunge of the sample $\mathbf{X}_1$ axis can be used for ϕ and λ, respectively in equations 3.2 and 3.3. Second, samples are often taken from geologic units that are no longer in the same position as when they were magnetized; they are tilted. If paleo-horizontal can be recognized, for example, from quasi-horizontal laminations in sedimentary rocks, the orientation of the bedding plane can be measured as *strike* and *dip*, or as dip and *dip direction*. The strike is the direction of a horizontal line within the bedding plane and the dip is the angle that the plane makes with the horizontal. Our convention is that dip is measured to the "right" of the strike direction. If the direction cosines relating the dip and dip direction to the geographic coordinate systems are plugged in for the a_{ij}, the data can be transformed into so-called *tilt adjusted* coordinates using equation 3.3.

[See Example 3.2]

3.3. Field strategies

In addition to establishing that a given rock unit retains a consistent magnetization, it is also of interest to establish when this magnetization was

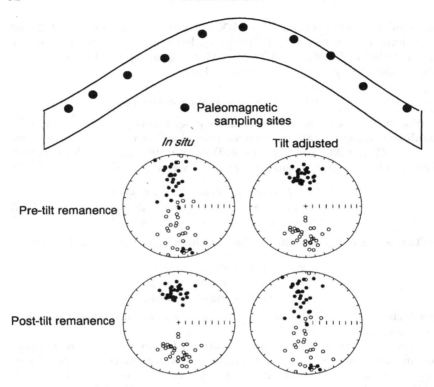

Figure 3.6.   Sampling units with different bedding attitudes in the "fold test". The paleomagnetic directions are shown on equal area projections before and after adjusting for bedding tilt (conventions are the same as for Figure 1.9).

acquired. Arguments concerning the age of magnetic remanence can be built on indirect petrographic evidence as to the relative ages of various magnetic minerals, or by evidence based on geometric relationships in the field. There are two key field tests that require special sampling strategies: the fold test and the conglomerate test.

The *fold test* relies on the tilting or folding of the target geological material. If, for example, one wanted to establish the antiquity of a particular set of directions, one could deliberately sample units of like lithology, with different present attitudes (Figure 3.6). If the recovered directions are more tightly grouped before adjusting for tilt (as in the lower left panel), then the magnetization is likely to have been acquired after tilting. On the other hand, if directions become better grouped in the tilt adjusted coordinates (see upper right panel), one has an argument in favor of a pre-tilt age of the magnetization. Methods for quantifying the tightness of grouping in various coordinate systems are discussed in Chapter 4 (see section on the "fold test").

In the *conglomerate test*, lithologies that are desirable for paleomagnetic purposes must be found in a conglomerate bed (Figure 3.7). In this rare and happy circumstance, we can sample them and show that: 1) the rock magnetic behavior is the same for the conglomerate samples as for those being used in the paleomagnetic study, 2) the directions of the studied lithology are well grouped, (Figure 3.7) and 3) the directions from the conglomerate clasts are randomly oriented (see Figure 3.7 and Chapter 4). If the directions of the clasts are not randomly distributed (Figure 3.7), then presumably the conglomerate clasts (and, by inference, the paleomagnetic samples from the studied lithology as well) were magnetized after deposition of the conglomerate. We will discuss statistical methods for deciding if a set of directions is random in Chapter 4.

The *baked contact test* is illustrated in Figure 3.8. It is similar to the conglomerate test in that we seek to determine whether the lithology in question has undergone pervasive secondary overprinting. When an igneous body intrudes into an existing *host rock*, it heats (or bakes) the contact zone to above the Curie temperature of the host rock. The baked contact immediately adjacent to the intrusion should therefore have the same remanence direction as the intrusive unit. This magnetization may be in an entirely different direction from the pre-existing host rock. The maximum temperature reached in the baked zone decreases away from the intrusion and remagnetization is not complete. Thus the NRM directions of the baked zone gradually change from that of the intrusion to that of the host rock. Such a condition would argue against pervasive overprinting in the host rock that post-dated the intrusion, and the age of the intrusion would provide an upper bound on the age of remanence in the host rock.

3.4. Measurement

We measure the magnetic remanence of paleomagnetic samples in a *magnetometer*, of which there are various types. The cheapest and most readily available are called *spinner magnetometers* because they spin the sample to create a fluctuating electromotive force (emf). The emf is proportional to the magnetization and can be determined relative to the three axes defined by the sample coordinate system. The magnetization along a given axis is measured by detecting the voltages induced by the spinning magnetic moment within a set of pick-up coils.

Another popular way to measure the magnetization of a sample is to use a *cryogenic magnetometer*. These magnetometers operate using so-called

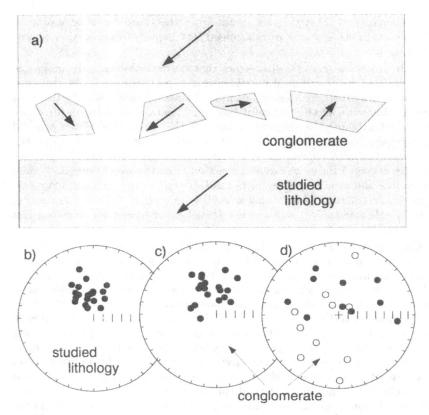

Figure 3.7. a) The paleomagnetic conglomerate test. The directions of samples from the studied lithology are shown in b) indicating that it is relatively homogeneously magnetized. Material from the studied lithology was incorporated into a conglomerate bed. If the conglomerate clasts are also homogeneously magnetized as in c) then the magnetization must post-date formation of the conglomerate. In a positive conglomerate test d), the magnetization vectors of samples from the conglomerate bed are random from clast to clast.

superconducting quantum interference devices (SQUIDs). In a SQUID, the flux of an inserted sample is opposed by a current in a loop of superconducting wire. The superconducting loop is constructed with a *weak link* which stops superconducting at some very low current density, corresponding to some very small quantum of flux. Thus the flux within the loop can change by discrete quanta. Each incremental change is counted and the total flux is proportional to the magnetization along the axis of the SQUID.

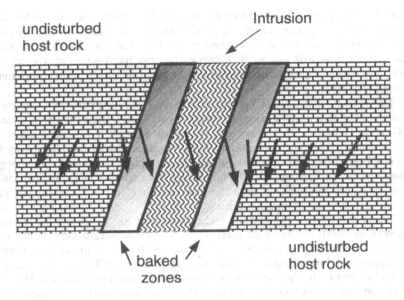

Figure 3.8. The baked contact test. In a positive test, zones baked by the intrusion are remagnetized and have directions that grade from that of the intrusion to that of the host rock. If all the material is homogeneously magnetized, then the age of the intrusion places an upper bound on the age of magnetization.

Cryogenic magnetometers are much faster and more sensitive than spinner magnetometers, but they cost much more to buy and to operate.

Magnetometers are used to measure the three components of the magnetization necessary to define a vector (e.g., x_1, x_2, x_3). These data can be converted to the more common form of D, I and M by methods described in Chapter 1.

3.5. Demagnetization techniques

Anyone who has dealt with magnets (including magnetic tape, credit cards, and magnets) knows that they are delicate and likely to demagnetize or change their magnetic properties if abused by heat or stress. Cassette tapes left on the dashboard of the car in the hot sun never sound the same. Credit cards that have been through the dryer may lead to acute embarrassment at the check-out counter. Magnets that have been dropped, do not work as well afterwards. It is not difficult to imagine that rocks that have been left in the hot sun or buried deep in the crust (not to mention altered by diagenesis or bashed with hammers, drills, pick axes, etc.), may not have their original magnetic vectors completely intact. Because rocks often contain millions of tiny magnets, it is possible that some (or all) of these have become realigned, or that they grew since the rock formed. In many cases, there are still grains that carry the original remanent vector, but there are often populations of grains that have acquired new components of magnetization.

Through geologic time, certain grains may acquire sufficient energy to overcome the magnetic anisotropy energy and change their direction of magnetization (Chapter 2). In this way, rocks can acquire a viscous magnetization in the direction of the ambient field. Because the grains that carry the viscous magnetization necessarily have lower magnetic anisotropy energies (they are "softer", magnetically speaking), we expect their contribution to be more easily randomized than the more stable ("harder") grains carrying the ancient remanent magnetization.

There are several laboratory techniques that are available for separating various components of magnetization. Paleomagnetists rely on the relationship of relaxation time, coercivity, and temperature in order to remove (*demagnetize*) low stability remanence components (Chapter 2). The fundamental principle that underlies demagnetization techniques is that the lower the relaxation time τ, the more likely the grain will carry a secondary magnetization. The basis for *alternating field* (AF) demagnetization is that components with short relaxation times also have low coercivities. The basis for *thermal* demagnetization is that these grains also have low blocking temperatures.

In AF demagnetization, an oscillating field is applied to a paleomagnetic sample in a null magnetic field environment. All the grain moments with coercivities below the peak AF will track the field. These entrained moments will become stuck as the peak field gradually decays below the coercivities of individual grains. Assuming that there is a range of coercivities in the sample, the low stability grains will be stuck half along one direction of the AF and half along the other direction; the net contribution to the remanence will be zero. In practice, we demagnetize samples sequentially along three orthogonal axes, or while "tumbling" the sample around three axes during demagnetization.

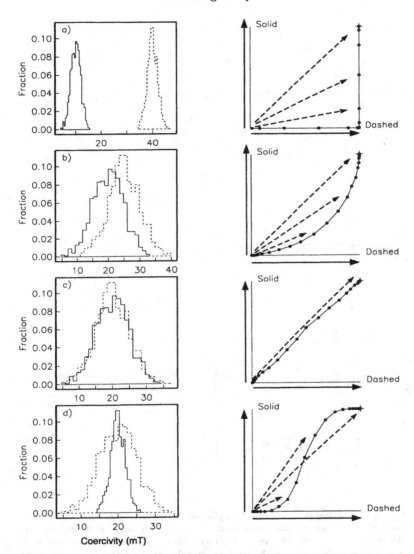

Figure 3.9. Principle of progressive demagnetization. Specimens with two components of magnetization (shown by heavy arrows on the right hand side), with discrete coercivities (plotted as histograms to the left). The original "NRM" is the sum of the two magnetic components and is shown as the + in the diagrams to the right. Successive demagnetization steps remove the component with coercivities lower than the peak field, and the NRM vector changes as a result. a) The two distributions of coercivity are completely separate. b) The two distributions partially overlap resulting in simultaneous removal of both components. c) The two distributions completely overlap. d) One distribution envelopes the other.

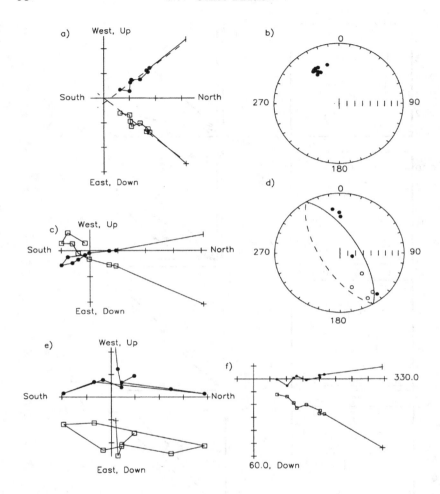

Figure 3.10. a) Orthogonal projection of data plotted with North on the horizontal axis. A single component of magnetization is present (see text). The horizontal projection is plotted with solid symbols and the vertical (North-down) component is plotted with open squares. The best-fit line through the data, as calculated by principal component analysis, is shown as a dashed line. b) Same data as in a), but plotted on an equal area projection. Solid (open) symbols are projections onto the lower (upper) hemisphere. c) Sample with two components that have overlapping stabilities. d) Same data as in c), but in an equal area projection. The trace of the best-fitting plane is shown, with a solid (dashed) line being a projection onto the lower (upper) hemisphere. e) Example of a complicated multi-component sample. f) Data from a) replotted with the horizontal axis along 330° instead of with respect to North.

We sketch the principles of progressive demagnetization in Figure 3.9. Initially, the NRM is the sum of two components carried by populations with different coercivities. The distributions of coercivities are shown in the histograms to the left in Figure 3.9. Two components of magnetization are shown as heavy lines in the plots to the right. In these examples, the two components are orthogonal. The sum of the two components at the start (the NRM) is shown as a + on the vector plots to the right. After the first AF demagnetization step, the contribution of the lowest coercivity grains has been erased and the remanence vector moves to the position of the first dot away from the +. Increasing the AF gradually eats away at the remanence vectors (shown as dashed arrows and dots in the plots to the right) which eventually approach the origin.

There are four different sets of coercivity spectra shown in Figure 3.9, each with a distinctive behavior during demagnetization. If the two co-ercivity fractions are completely distinct, the two components are clearly defined (Figure 3.9a) by the progressive demagnetization. Hoffman and Day [1978] (see also Zijderveld [1967]) pointed out however, that if there is some overlap in the coercivity distribution of the components the resulting de-magnetization diagram is curved (Figure 3.9b). If the two components completely overlap, both components are removed simultaneously and an apparently single component demagnetization diagram may result (Fig-ure 3.9c). It is also possible for one coercivity spectrum to include another as shown in Figure 3.9d. Such cases result in "S" shaped demagnetization curves.

Because complete overlap actually happens in "real" rocks, it is desirable to perform both AF and thermal demagnetization. If the two components overlap completely in coercivity, they might not have overlapping blocking temperature distributions and vice versa. It is unlikely that samples from the same lithology will all have identical overlapping distributions, so mul-tiple samples can provide clues to the possibility of completely overlapped directions in a given sample.

3.6. Demagnetization data display

The standard practice in demagnetization is to measure the NRM and then to subject the sample to a series of demagnetization steps of increas-ing severity. The magnetization of the sample is measured after each step. During demagnetization, the remanent magnetization vector will change until the most stable component has been isolated, at which point the vec-tor decays in a straight line to the origin. This final component is called the *characteristic remanent magnetization* or ChRM. Visualizing demagnetiza-tion data is a three-dimensional problem and therefore difficult to plot on paper. Paleomagnetists often rely on a set of two projections of the vectors, one on the horizontal plane and one on the vertical plane. These are variously called Zijderveld diagrams (Zijderveld [1967]), orthogonal projections, or vector end-point diagrams.

In orthogonal projections, the North component (x_1) is plotted versus East (x_2) (solid symbols) in one projection, and North (x_1) is replotted versus Down (x_3) (open symbols) in another projection. Here, paleomagnetic convention differs from the usual x-y plotting convention because x_2 and x_3 are on the $-y$ axis. The paleomagnetic conventions make sense if one visualizes the diagram as a map view for the solid symbols and a vertical projection for the open symbols. It may be advantageous to plot North on the vertical axis and East positive to the right. In this case the vertical projection is East versus Down. This projection is useful if the magnetization is more East-West than North-South. In fact, the horizontal axis can be any direction within the horizontal plane.

In Figure 3.10, we show three general types of demagnetization behavior. In Figure 3.10a-b, the sample has a North-Northwest and downward directed NRM (plotted as +'s). The direction does not change during demagnetization and the NRM is a single vector. The directional data are also plotted on the equal area net to the right (Figure 3.10b) and fall in the NW quadrant of the lower hemisphere. The sample in Figure 3.10c shows a progressive change in direction from a North-Northwest and downward directed component to a South-Southeast and upward direction. The vector continuously changes direction to the end and no final "clean" direction has been confidently isolated. These data are plotted on an equal area projection to the right (Figure 3.10d) along with the trace of the best-fitting plane (a great circle). The most stable component probably lies somewhere near the best-fitting plane.

In Figure 3.10e, we show what is informally known as a "spaghetti" diagram. The NRM switches from direction to direction, with little coherence from step to step. Such data are difficult to interpret and are usually thrown out.

Some people choose to plot the pairs of points (x_1, x_2) versus (H, x_3) where H is the horizontal projection of the vector given by $\sqrt{x_1^2 + x_2^2}$. In this projection, which is sometimes called a *component plot*, the two axes do not correspond to the same vector from point to point. Instead, the coordinate system changes with every demagnetization step because H almost always changes direction, even if only slightly. Plotting H versus x_3 is therefore a confusing and misleading practice. The primary rationale for doing so is because, in the traditional orthogonal projection, the vertical component reveals only an apparent inclination. If something close to true inclination is desired, then, instead of plotting H and x_3, one can simply rotate the horizontal axes of the orthogonal plot such that it closely parallels the desired declination (Figure 3.10f).

3.7. Vector difference sum

An equal area projection may be the most useful way to present demagnetization data from a sample with several strongly overlapping remanence components (such as in Figures 3.10c-d and 3.11). In order to represent the vector nature of paleomagnetic data, it is necessary to plot intensity information. Intensity can be plotted versus demagnetization step in an *intensity decay curve* (Figure 3.11c). However, if there are several components with different directions, the intensity decay curve cannot be used to determine, say, the blocking temperature spectrum, because it is the vector sum of the two components. It is therefore advantageous to consider the decay curve of the *vector difference sum* (VDS.) The VDS "straightens out" the various components by summing up the vector differences at each demagnetization step, so the total magnetization is plotted, as opposed to the resultant (see Figure 3.11).

3.8. Principal component analysis

Orthogonal vector projections aid in identification of the various remanence components in a sample. The remanence directions are usually calculated by *principal component analysis* (Kirschvink [1980]). A sequence of data points which form a single component are equally weighted. The D, I, and M data are converted to corresponding x values (see Chapter 1). Then we calculate the coordinates of the "center of mass" ($\bar{x}$) of the data points:

$$\bar{x}_1 = \frac{1}{N}(\sum_1^N x_{1i}); \quad \bar{x}_2 = \frac{1}{N}(\sum_1^N x_{2i}); \quad \bar{x}_3 = \frac{1}{N}(\sum_1^N x_{3i}), \qquad (3.4)$$

where N is the number of data points involved. We then transform the origin of the data cluster to the center of mass:

$$x'_{1i} = x_{1i} - \bar{x}_1; \quad x'_{2i} = x_{2i} - \bar{x}_2; \quad x'_{3i} = x_{3i} - \bar{x}_3, \qquad (3.5)$$

where x'_i are the transformed coordinates.

3.8.1. THE ORIENTATION TENSOR AND EIGENVECTOR ANALYSIS

The *orientation tensor* **T** (Scheidegger [1965]) (also known as the matrix of sums of squares and products), is extremely useful in paleomagnetism:

$$\mathbf{T} = \begin{pmatrix} \sum x'_{1i}x'_{1i} & \sum x'_{1i}x_{2i} & \sum x'_{1i}x'_{3i} \\ \sum x'_{1i}x'_{2i} & \sum x'_{2i}x'_{2i} & \sum x'_{2i}x'_{3i} \\ \sum x'_{1i}x'_{3i} & \sum x'_{2i}x'_{3i} & \sum x'_{3i}x'_{3i} \end{pmatrix}. \tag{3.6}$$

$\mathbf{T}$ is a 3 x 3 matrix, where only six of the nine elements are independent. It is constructed in some coordinate system, such as the geographic or sample coordinate system. Usually, none of the six independent elements are zero. There exists, however, a coordinate system along which the "off-axis" terms are zero and the axes of this coordinate system are called the *eigenvectors* of the matrix. The three elements of $\mathbf{T}$ in the eigenvector coordinate system are called *eigenvalues*. In terms of linear algebra, this idea can be expressed as:

$$\mathbf{TV} = \tau \mathbf{V}, \tag{3.7}$$

where $\mathbf{V}$ is the matrix containing three *eigenvectors* and τ is the diagonal matrix containing three *eigenvalues*. Equation 3.7 is only true if:

$$\det|\mathbf{T} - \tau| = 0. \tag{3.8}$$

If equation 3.8 is expanded, we have a third degree polynomial whose roots (τ) are the eigenvalues:

$$(T_{11} - \tau)[(T_{22} - \tau)(T_{33} - \tau) - T_{23}^2] -$$

$$T_{12}[T_{12}(T_{33} - \tau) - T_{13}T_{23}] + T_{13}[T_{13}T_{23} - T_{13}(T_{22} - \tau)] = 0.$$

The three possible values of τ (τ_1, τ_2, τ_3) can be found with iteration and determination. In practice, there are many programs for calculating τ and the reader is referred to Press et al. [1986] for a thorough discussion. Please note that the conventions adopted here are to scale the τ's such that they sum to one; the largest eigenvalue is termed τ_1 and corresponds to the eigenvector $\mathbf{V}_1$.

3.8.2. PRINCIPAL COMPONENTS OF THE ORIENTATION MATRIX

Inserting the values for the transformed components calculated in equation 3.5 into $\mathbf{T}$ gives the covariance matrix for the demagnetization data. The direction of the axis associated with the greatest scatter in the data (the principal eigenvector $\mathbf{V}_1$) corresponds to a best-fit line through the data. This is usually taken to be the direction of the component in question. This direction also corresponds to the axis around which the "moment of inertia" is least. The eigenvalues of $\mathbf{T}$ are the variances associated with each eigenvector. Thus the standard deviations are $\sigma_i = \sqrt{\tau_i}$. The so-called *maximum angular deviation* or MAD of Kirschvink [1980] is defined as:

$$MAD = \tan^{-1}(\sqrt{(\sigma_2^2 + \sigma_3^2)}/\sigma_1).\qquad(3.9)$$

[See Example 3.3]

If no unique principal direction can be isolated (as for the sample in Figure 3.10c-d), the eigenvector V_3 associated with the least eigenvalue τ_3 can be taken as the pole to the best-fit plane wherein the component of interest must lie. Kirschvink [1980] also defines a MAD angle for the plane as:

$$MAD_{\text{plane}} = \tan^{-1}\sqrt{\tau_3/\tau_2 + \tau_3/\tau_1}.\qquad(3.10)$$

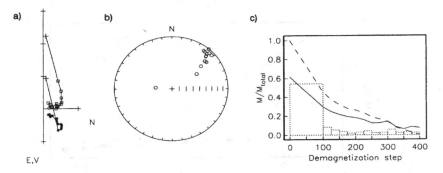

Figure 3.11. a) Specimen with strongly overlapping remanence components, in an orthogonal projection. b) Same data as in a) plotted on an equal area projection. c) Decay of NRM intensity during the demagnetization procedure (solid line). The dashed line is the decay of the vector difference sum. Boxes represent the intensity removed after each step.

3.9. Blocking temperature and coercivity spectra

Because geological materials are often complicated mixtures of several magnetic minerals, it is useful to determine the coercivity and/or blocking temperature spectra of the NRM using the VDS technique described in the last section. By plotting the VDS curves for sister samples that underwent AF and thermal demagnetization, much can be learned about the carrier the NRM.

Some important magnetic phases in geological materials (Table 2.1) are magnetite (maximum blocking temperature of ~580°C, maximum coercivity of about 0.3 T), hematite (maximum blocking temperature of ~ 675°C and maximum coercivity much larger than 5 T), goethite (maximum blocking temperature of ~ 125°C and maximum coercivity of much larger than 5

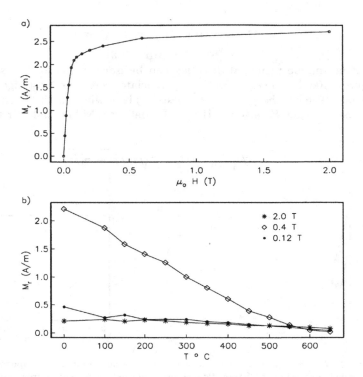

Figure 3.12. a) Acquisition of IRM (M_r). After applying a field of 2 T, the sample was subjected to two additional IRM's: 0.4 T and 0.12 T along orthogonal axes. b) Thermal demagnetization of a 3-axis IRM. Each component is plotted separately.

T), and various sulfides. The relative importance of these minerals in bulk samples can be constrained by a simple trick that exploits both differences in coercivity and unblocking temperature (Lowrie [1990]).

The Lowrie "three axis IRM test" proceeds as follows:

• Apply an IRM along three orthogonal directions in three different fields. The first field, applied along X_1, should be sufficient to saturate all the minerals within the sample and is usually the largest field achievable in the laboratory (say 2 T). The second field, applied along X_2, should be sufficient to saturate magnetite, but not to realign high coercivity phases, such as goethite or fine-grained hematite (say 0.4 T). The third IRM, applied along X_3, should target low coercivity minerals and the field chosen is typically something like 0.12 T.

• The composite magnetization can be characterized by determining the blocking temperature spectra for each component. This is done by thermally demagnetizing the sample and plotting the magnitude of the x_1, x_2, and x_3 components versus demagnetizing temperature.

[See Example 3.4]

An example of three axis IRM data are shown in Figure 3.12. The curve is dominated by a mineral with a maximum blocking temperature of between 550° and 600°C and has a coercivity less than 0.4 T, but greater than 0.12 T. These properties are typical of magnetite (Table 2.1). There is a small fraction of a high coercivity (>0.4 T) mineral with a maximum unblocking temperature > 650°C, which is consistent with the presence of hematite (Table 2.1).

3.10. Paleointensity determination

We turn now to the question of the variation in intensity of the Earth's magnetic field. Until now we have only discussed methods for obtaining directional data from rock samples. In principle, it is also possible to determine the intensity of ancient magnetic fields, because the primary mechanisms by which rocks become magnetized (e.g., thermal, chemical and detrital remanent magnetizations) are approximately linearly related to the ambient field (for low fields such as the Earth's). In the ideal case, one would determine the proportionality factor that relates magnetization and field. For example, because TRM is proportional to the magnitude of **B**, we need only measure the NRM, then give the rock a TRM in a known field. Because

$$\frac{TRM_{lab}}{B_{lab}} = \frac{NRM}{B_{ancient}},$$

the ambient field responsible for the NRM ($B_{ancient}$) is easily calculated. Similarly, for a DRM, one could redeposit the sediment in a known field and thereby calculate the ratio of DRM to B and estimate the ancient field from the NRM. In practice, however, there are problems which limit the usefulness of each of these simple approaches. We discuss more appropriate approaches for obtaining paleointensity information from TRMs and DRMs in the following sub-sections.

3.10.1. PALEOINTENSITY WITH TRMS

The simple method for extracting paleointensity data from TRMs outlined in the foregoing text often fails because rocks alter when heated in the laboratory. The TRM that is acquired by heating the rock to above its Curie temperature and cooling in a known field may have no relationship

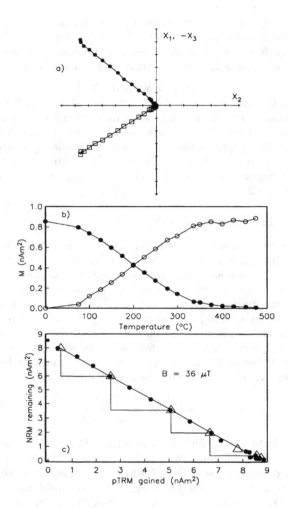

Figure 3.13. Illustration of the Thellier-Thellier method for determining absolute paleointensity (redrawn from Pick and Tauxe [1993]): a) an orthogonal plot of stepwise demagnetized NRM data in sample coordinates, b) thermal demagnetization of NRM shown as filled circles and the laboratory acquired pTRM shown as open symbols, and c) Arai plot of NRM component remaining versus pTRM acquired for each demagnetization temperature. So-called "pTRM checks" are triangles.

to the original TRM. Some minerals, such as maghemite, are unstable at elevated temperatures and convert to other magnetic phases. Magnetite also oxidizes at elevated temperatures. The NRM may have a component of VRM as well as TRM. Such changes in the magnetic properties can have a drastic effect on the magnetization. We therefore need a way to detect such problems and to compensate for them.

The most effective way to determine paleointensity from igneous rocks was suggested by Thellier and Thellier [1959] and was refined by Coe et al. [1967a,b] (see also Coe et al. [1978], Prévot et al. [1985], and Pick and Tauxe [1993]). The Thellier-Thellier technique relies on a property of single-domain TRM called the law of additivity of partial thermal remanence (Chapter 2). This law states that if a sample is heated to below the Curie temperature, only grains whose blocking temperatures are below that temperature will be affected, so the pTRM that is acquired when cooled in a field does not affect what remains of the original TRM. In practice, the relationship between the TRM and the field can be obtained by first measuring the NRM, then by heating the sample to some temperature, say 100°C, and cooling the sample in zero field. The NRM is then remeasured. The sample is then reheated to the same temperature but is cooled in a known field (of the order of the Earth's field) and measured again. This double heating procedure is repeated until the highest unblocking temperature is reached. The reproducibility of the pTRM step can be assessed by performing a so-called *pTRM check*, whereby a pTRM at some lower temperature is re-introduced after the NRM measurement at a higher temperature step. If the second pTRM is significantly different from the first, then the capacity of the sample to acquire TRM has changed, which provides a strong hint that the pTRM data acquired above that temperature are unreliable.

[See Example 3.5]

In Figure 3.13 we show results of a Thellier-Thellier experiment. The data can be plotted in a number of ways; the *Arai* plot of Nagata et al. [1963] is perhaps the most popular (see Figure 3.13c). In the Arai plot, the pTRM gained in the "in-field" step is plotted versus the NRM remaining after the "zero field" step at the same temperature. If a sample does not alter during the experiment, a single component has been isolated, and the law of additivity of pTRM holds true. In this case, the NRM-pTRM ratio should remain constant throughout the experiment. Thus, the ideal Arai plot results in a straight line with a negative slope. Provided that the TRM is linearly related to the applied field, the absolute value of the slope of the line equals the ratio of ancient field to laboratory field. Deviations from a line result from multi-domain behavior (e.g., Dunlop and Xu [1994]), sample alteration, and/or from non-ideal experimental conditions such as substantial differences in cooling rate, or poorly controlled temperature or sample orientation.

The procedure for calculating the best-fit slope, which is the best estimate for the paleofield, is as follows:

- Taking the N data points that span two temperature steps T_1 and T_2,
the best-fit slope b relating the NRM (y_i) and the pTRM (x_i) data in a
least squares sense (taking into account variations in both x and y (see
Coe et al. [1978]) is given by:

$$b = \sqrt{\frac{\sum_i (y_i - \bar{y})^2}{\sum_i (x_i - \bar{x})^2}}, \tag{3.11}$$

where $\bar{y}$ is the average of all y values and $\bar{x}$ is the average of all x values.

- The standard error of the slope σ is:

$$\sigma = \sqrt{\frac{\sum_i (y_i - \bar{y})^2 - 2b \sum_i (x_i - \bar{x})(y_i - \bar{y})}{(N - 2) \sum_i (x_i - \bar{x})^2}}. \tag{3.12}$$

[See Example 6.4]

- The data can be iteratively searched to find the portion of data with
the lowest value of $\sigma/|b|$ which is a measure of the uncertainty in the
slope (see Coe et al. [1978]).

- The paleofield strength $B_{ancient} = |b|B_{lab}$, and the equivalent moment
of the axial dipole is given by equation 1.20 from Chapter 1.

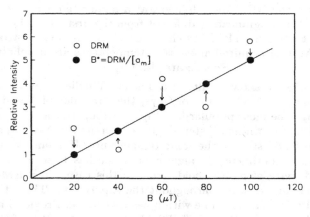

Figure 3.14. Principles of relative paleointensity. The original DRM is plotted as open
symbols. It is a function not only of the applied field, but also of the magnetic activity
$[a_m]$ of the sample. When normalized by $[a_m]$ (dots), the DRM is a linear function of
applied field B.

3.10.2. PALEOINTENSITY WITH DRMS

The principle on which paleointensity studies in sedimentary rocks rests is that DRM is linearly related to the magnitude of the applied field **B** under certain circumstances in the laboratory (e.g., Kent [1973]). The problem with sedimentary paleointensity data is that laboratory conditions cannot duplicate the natural environment. First, most sediments carry a post-depositional remanence (or chemical remanence) as opposed to a depositional remanence (see Tauxe [1993] for a thorough review). Second, the intensity of remanence is a function not only of field but of magnetic mineralogy and even chemistry of the water column (e.g., Lu et al. [1990]). Third, sedimentary rocks are lithified and cannot easily be disaggregated to their original state.

Under the ideal conditions depicted in Figure 3.14, the initial DRM of a set of samples deposited under a range of magnetic field intensities (B) is shown as open circles. The relationship is not linear because each sample has a different *magnetic activity* $[a_m]$ as a result of differences in the amount of magnetic material, magnetic mineralogy, etc. For example, samples with a higher concentration of magnetic material will have a higher DRM. If $[a_m]$ can be successfully approximated, for example, by bulk remanences such as IRM or ARM, or by bulk magnetic susceptibility χ_b (Chapter 2), then a normalized DRM (shown as dots in Figure 3.14) will reflect the relative intensity of the applied field.

Our theoretical understanding of DRM is much less developed than for TRM (Chapter 2). Because of the lack of a firm theoretical foundation for DRM, there is no simple method for determining $[a_m]$. Many proxies have been proposed (see review by Tauxe [1993]) ranging from normalization by bulk magnetic properties such as ARM, IRM, or χ, to Thellier-Thellier type normalization (Tauxe et al. [1995], Hartl and Tauxe [1996b]). The principal assumption in the Thellier-Thellier type experiment in sediments is that the DRM component behaves in the same way during thermal demagnetization as would a TRM in the same assemblage (i.e., DRM is linear with TRM). If this is true for a particular sedimentary sample, the data acquired during a Thellier-Thellier experiment on the sample will yield a linear Arai plot (Hartl and Tauxe [1996]).

Sediments are not always amenable to thermal demagnetization, and it is possible to modify the Thellier-Thellier experiment by using the acquisition of pARMs in place of the pTRMs in the traditional Thellier-Thellier experiment, in the so-called *pseudo-Thellier* experiment (Tauxe et al. [1995]). In the pseudo-Thellier experiment (Figure 3.15), the NRM is first AF demagnetized in a stepwise fashion (Figure 3.15a and stars in Figure 3.15b). Then, pARMs are imparted at the same AF steps (solid squares in Figure

3.15b). When the pARMs are plotted against the remaining NRM in an Arai plot (Figure 3.15c), the slope can be calculated in the same manner as for Thellier-Thellier data. The pseudo-Thellier experiment may be appropriate for sediments that alter or disintegrate on heating. Thellier type experiments serve well to normalize sedimentary paleointensity in a relative sense and have the advantage of efficiently separating VRM from DRM (see Kok and Tauxe [1996]). Because these experiments are extremely time consuming, it may be sufficient to do them on a selection of pilot samples. These data can be used to demonstrate the optimum "blanket" treatment, which would then be applied to the rest of the samples.

[See Example 3.6]

The best for which we can presently hope from sediments is relative paleointensity (but see Constable and Tauxe [1996]). Nonetheless, there is cause for optimism that relative paleointensities can be reliably determined (see King et al. [1983], Tauxe [1993], Guyodo and Valet [1996], Frank et al. [1996], and Constable et al. [1998]).

3.11. Summary

In this chapter, we have outlined how to obtain samples and reviewed routine laboratory analyses. The end product of these efforts is a set of paleomagnetic vectors which one hopes represent ancient geomagnetic field conditions. Calculation of mean directions and quantification of scatter of paleomagnetic vector data is the subject of the following chapter.

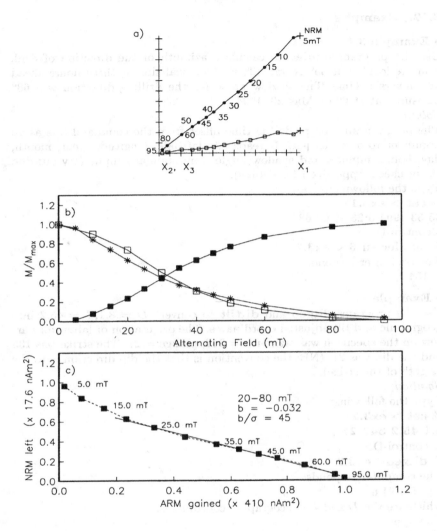

Figure 3.15. Illustration of the pseudo-Thellier technique. a) Vector end-point diagram of stepwise AF demagnetization data. X_1 and X_2 (solid symbols) are in the horizontal plane unoriented with respect to North, and X_1 and X_3 (open symbols) are in the vertical plane. Plus symbols mark the NRM data. b) Remanence normalized by maximum values for NRM (stars), and ARM (squares). The solid (open) squares are the acquisition (demagnetization) of ARM. c) Arai plot for the data in b). The pARM gained in specified alternating fields versus NRM remaining after demagnetization in the same field.

3.12. Examples

• Example 3.1

Use the program **sundec** to calculate azimuth of the direction of drill. You are located at 35° N and 33° E. The local time is three hours ahead of Universal Time. The shadow angle for the drilling direction was 68° measured at 16:09 on May 23, 1994.

Solution

The program **sundec** reads the time offset from the command line as an argument to a switch [**-u**]. It reads site latitude, longitude, year, month, day, hours, minutes and shadow angle from standard input (try **sundec -h**, or check Appendix 1 for details).

Type the following:

% **cat > ex3.1**
35 33 1994 5 23 16 09 68
<control-D>
% **sundec -u 3 < ex3.1**
The computer responds:
    154.2

• Example 3.2

Use the programs **di_geo** and **di_tilt** to convert $D = 8.1, I = 45.2$ into geographic and tilt adjusted coordinates. The orientation of laboratory arrow on the specimen was: azimuth = 347; plunge = 27. The strike was 135 and the dip was 21. (NB: the convention is that the dip direction is to the "right" of the strike).

Solution

Type the following:

% **cat > ex3.2**
8.1 45.2 347 27
< control-D>
% **di_geo< ex3.2**
The computer responds:
    5.3 71.6
which are the D and I in geographic coordinates.
Now type:
% **cat > ex3.2a**
5.3 71.6 135 21
< control-D>
% **di_tilt< ex3.2a**
The computer responds:
    285.7 76.6
which are the D and I in tilt adjusted coordinates.

- **Example 3.3**

1) Use the program **pca** to calculate the best-fit direction from the data in Table 3.1. Use only the 20 to 90 mT demagnetization steps. 2) Use the program **gtcirc** to calculate the best-fit plane through the NRM to 20 mT steps. 3) Plot the data (and the best-fit line) using the program **plotdmag**. 4) Plot the data with $D = 300$ on the $\mathbf{X}_1$ axis.

Solution

Because each laboratory has its own file formatting convention, it is necessary to extract the desired data from a given file format using the **grep** and **awk** commands. If the above data are in a file named **yourdata** along with data from other samples, then the data for the particular sample **eba24a** can be extracted using the **grep** command. Type:

% **grep "eba24a" yourdata > tmp**

Three data formats are supported by **plotdmag** (Appendix 1). If the data format of your file is different, the columns needed by **plotdmag** can be stripped out using **awk**. For example, if columns 1, 3, 5, 6, and 7 are required, then type the following:

% **awk '{print $1,$3,$5,$6,$7}' tmp > ex3.3**

These two commands can be combined:

% **grep "eba24a" yourdata | awk '{print $1, $3,$5,$6,$7}' > ex3.3**

Now **pca** can be made to analyze the data from the 6th (20 mT) to the 13th (90 mT) treatment steps as follows.

Type:

% **pca -p 6 13 < ex3.3**

The computer responds:

   eba24a p 8 20.00 90.00 4.5 307.9 39.8

To find out what these numbers mean, check Appendix 1 or type:

% **pca -h**

and get:

Usage pca [-pmd] [beg end][ta][Standard I/O]
     calculates best-fit line through specified input data
       -p PCA from [beg] to [end] steps
       -d uses .dat file as input
       if [ta]=0 (default), uses geographic (fdec,finc)
       if [ta] = 1 uses tilt adjusted (bdec,binc)
      -m uses .mag file as input
    Input options:
    Default input:
      Sample name tr int dec inc
        .mag file option
      Sample name tr csd int dec inc
        .dat file option

Sample name pos tr csd int fdec finc bdec binc

Output is:

Sample name p n beg end mad dec inc

where dec and inc are for the princ. comp.

The "**p**" indicates that this is a principal component direction (as opposed to a pole to a great circle), **n** is the number of data points used, [**beg**] and [**end**] are the bounding treatment steps used, **mad** is the MAD angle, and the declination and inclination are for the principal component. The output can be saved into a file using the UNIX redirect capability. To append to the end of a file called **pca.out**, type:

% **pca -p 6 13** < **ex3.3** >> **pca.out**

To calculate the pole to the best-fit plane and a corresponding MAD_{plane}, type the following:

% **gtcirc -g 1 6** < **ex3.3**

The computer responds:

eba24 g 6 0.00 20.00 7.4 193.1 29.3

This output is similar to that from **pca** except that the **g** stands for great circle and the direction is that of the pole to the best-fitting plane. Finally, to plot the data with the principal component along an axis with a declination of 300° and to view it with ghostview, type the following:

% **plotdmag -rp 300 6 13** < **ex3.3** | **plotxy ; ghostview mypost**

This causes plotxy to create a postscript file **mypost** as shown in Figure 3.16.

The program **plotdmag** can be used iteratively, first with no options for a quick look at the data and to display the index numbers for the treatment steps, then again to calculate best-fit lines and planes. To view a whole data file (e.g., **yourfile**), one can make a shell script as in the following example.

First list all the samples in the file (presuming that the sample name is in the first column.

% **awk '{print $1}' yourfile | sort | uniq > yourshell**

which makes a list of unique sample names (e.g., **s1, s2**, etc.). Now edit **yourshell** to look something like this:

grep s1 yourfile | awk '{print $1,$3,$5,$6,$7}' | plotdmag | plotxy;cat mypost > your.ps

grep s2 yourfile | awk '{print $1,$3,$5,$6,$7}' | plotdmag | plotxy;cat mypost >> your.ps

grep s3 yourfile | awk '{print $1,$3,$5,$6,$7}' | plotdmag | plotxy;cat mypost >> your.ps

grep s4 yourfile | awk '{print $1,$3,$5,$6,$7}' | plotdmag | plotxy;cat mypost >> your.ps

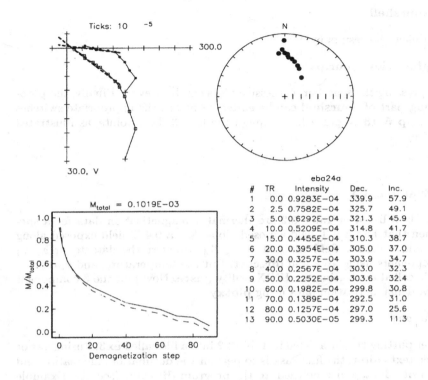

Figure 3.16. Output from Example 3.3, using the command: **plotdmag -rp 300 6 13** < **ex3.3** | **plotxy**. Top left is an orthogonal projection with the X_1 component aligned along $D = 300$. The horizontal (vertical) projection uses solid (open) symbols. The units for the tick marks are whatever the units of the data file are. Top right is an equal area projection of the same data. Lower left: the solid line is the intensity decay curve of the NRM data and the dashed line is the VDS decay curve. Lower right is a printout of the data with the index numbers of the demagnetization steps. The heavy dashed line in the orthogonal projection is the principal component calculated between steps #6 and # 13 which correspond to 20.0 and 90.0 mT respectively.

grep s5 yourfile | awk '{print \$1,\$3,\$5,\$6,\$7}' | plotdmag | plotxy;cat mypost >> your.ps

Now you must make **yourshell** executable, by e.g.:

% **chmod u+x yourshell**

Execute it by:

% **yourshell**

and view the results by:

% **ghostview your.ps**

By pressing the spacebar, successive plots can be viewed. Finally, the **plotdmag** part of **yourshell** can be edited to put in the appropriate switches (e.g., **-p 6 13** to put a line through the 6 - 13 data points, as illustrated above).

• **Example 3.4**

The data listed in Table 3.2 are thermal demagnetization data for a specimen that had a 2 T field exposed along X_1, a 0.4 T field exposed along X_2 and a 0.12 T field exposed along X_3. Convert the data to x_1, x_2, x_3 components using **awk** and **dir_cart**. Put the temperature, and x_i data in a file called **cart** using the UNIX utility **paste**. Now plot the 3-component IRM demagnetization data using **plotxy**.

Solution

After putting the data listed in Table 3.2 into a file called **ex3.4** using **cat** or some text editor, the first task is to peel off the declination, inclination and intensity data, and pipe them to the program **dir_cart** (see also Example 1.1). Put them in a file called **cart**. This is done with the command:

% **awk '{print \$2,\$3,\$4}' ex3.4 | dir_cart -m > cart**

Now we need to peel off the temperature steps from **ex3.4** using **awk** and paste them with the cartesian component data into a file **ex3.4a**.

% **awk '{print \$1}' ex3.4 > tmp; paste tmp cart > ex3.4a**

Finally we can plot the data using **plotxy** by putting the commands listed in Table 3.3 into a command file **ex3.4.com**. Notes as to the meaning of the commands are given to the right of Table 3.3.

Now type:

% **plotxy<ex3.4.com**

The output file **mypost** is shown in Figure 3.17.

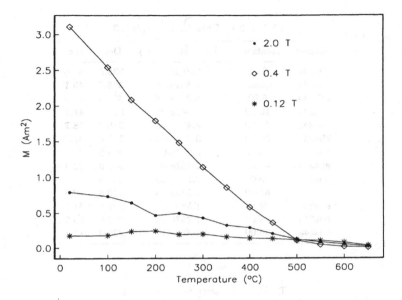

Figure 3.17. Output of thermal demagnetization data for a composite, 3-component IRM from Example 3.4. Plot generated by the following command: **awk '{print \$2,\$3,\$4}' ex3.4 | dir_cart -m > cart; awk {print \$1} ex3.4 > tmp; paste tmp cart > ex3.4a; plotxy < ex3.4.com.**

• **Example 3.5**
1) Use the program **arai** to convert the data from a paleointensity experiment contained in file **ex3.5** to an Arai plot. The laboratory field applied during the "in-field steps" was 50 μT. At which temperature step do the pTRM checks (triangles) fail? 2) Calculate a best-fit line between the 175° and 350°C steps.
Solution
First check Appendix 1 for our conventions. To make an Arai plot, calculating a best-fit line between the 175° and the 350°C temperature steps using a laboratory field of 50μT, type:
% arai -sf 175 350 50e-6 < ex3.5 | plotxy
The result is the **mypost** file shown in Figure 3.18.

• **Example 3.6**
Use the program **pseudot** to plot an Arai diagram for the pseudo-Thellier experiment whose data are contained in file **ex3.6**.
Solution

TABLE 3.1. Data for Example 3.3.

Sample	Treatment	Intensity (Am^2)	Dec.	Inc.
eba24a	0.00	9.283e-08	339.9	57.9
eba24a	2.50	7.582e-08	325.7	49.1
eba24a	5.00	6.292e-08	321.3	45.9
eba24a	10.00	5.209e-08	314.8	41.7
eba24a	15.00	4.455e-08	310.3	38.7
eba24a	20.00	3.954e-08	305.0	37.0
eba24a	30.00	3.257e-08	303.9	34.7
eba24a	40.00	2.567e-08	303.0	32.3
eba24a	50.00	2.252e-08	303.6	32.4
eba24a	60.00	1.982e-08	299.8	30.8
eba24a	70.00	1.389e-08	292.5	31.0
eba24a	80.00	1.257e-08	297.0	25.6
eba24a	90.00	5.03e-09	299.3	11.3

TABLE 3.2. Data for Example 3.4.

Treatment	Dec.	Inc	Intensity (Am^2)
20.00	75.7	3.1	3.21
100.00	73.9	3.8	2.65
150.00	72.8	6.1	2.19
200.00	75.4	7.4	1.86
250.00	71.5	6.9	1.57
300.00	69.4	9.1	1.23
350.00	69.5	9.6	0.924
400.00	63.8	11.8	0.659
450.10	60.7	17.5	0.426
500.00	41.7	34.9	0.192
550.00	28.1	48.3	0.134
600.00	17.6	54.9	0.09
650.00	18.9	56.8	0.0375

Type the following:
pseudot < ex3.6 | plotxy; ghostview mypost
and see Figure 3.19.

TABLE 3.3. Plotxy commands for Example 3.4.

Command	Comment
char .1	sets character size to 0.1 in
file ex3.4a	specifies the input file
frame	puts an attractive frame around the plot
affine 1 0 -1 0	multiplies x by 1, adds 0 to x
	multiplies y by -1, adds 0 to y
mode 20 1 2	specifies the columns (1 & 2) to be read as x & y
read	tells the program to read the file
symbol 19	changes the symbol from a line to a dot
read	tells the program to read the file
mode 20 1 3	specifies columns 1 & 3 as x & y
dash 0	changes the symbol back to a line
read	reads the data
symbol 3	changes the symbol to a diamond
read	reads the data
mode 20 1 4	specifies columns 1 & 4 as x & y
dash 0	changes the symbol back to a line
read	reads the data again
symbol 5	changes the symbol to an asterisc
read	reads the data again
xlabel Temperature	makes the X label
ylabel M	makes the Y label
xlimit 5 20 700	sets the X axis length to 5 inches
	and selects values between 20 and 700
ylimit 4 0 0	sets Y axis length to 4 inches and selects autoscaling
title Lowrie 3-component IRM	makes the plot title
plot 2 4	plots with origin offset by 2 and 4 inches in X and Y
stop	stops the program and writes the **mypost** file

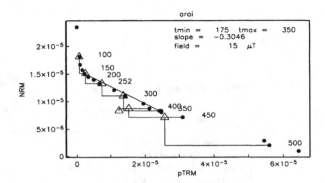

Figure 3.18. Output of Example 3.5 using the command: **arai -sf 175 350 50e-6 <
ex3.5 | plotxy.**

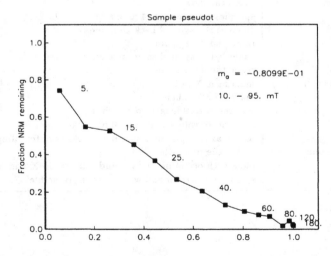

Figure 3.19. Output of Example 3.6 using the command: **pseudot < ex3.6 | plotxy.**

Chapter 4

ANALYZING VECTORS

In the previous chapter, we learned about routine paleomagnetic sampling and laboratory procedures. Once paleomagnetic directions have been obtained after stepwise demagnetization, principal component analysis, etc., one may wish to interpret them in terms of ancient geomagnetic field directions. To do this, there must be some way of calculating mean vectors and of quantifying the confidence intervals about a mean vector. In order to do this, we must understand the statistics of paleomagnetic directions.

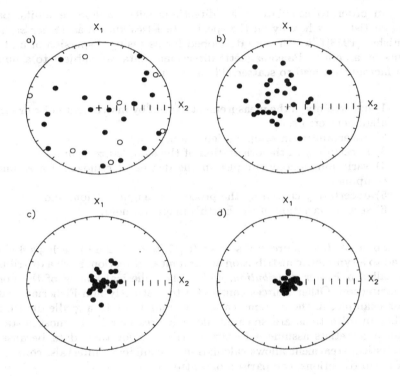

Figure 4.1. Four hypothetical data sets with decreasing scatter: a) is nearly uniformly distributed on the sphere, whereas d) is fairly well clustered. All data sets were drawn from Fisher distributions with vertical true directions.

Consider the various sets of directions that are shown in Figure 4.1, which are drawn from distributions with a vertical direction. It would be handy to be able to calculate a mean direction and to quantify the scatter of these directions. The average inclination, calculated as the sum of all the inclinations and divided by the number of points, will obviously not be vertical. We will see, however, that the average direction of each data set is actually nearly vertical. In the following, we will demonstrate the proper way to calculate mean directions and confidence regions for directional data that are distributed in the manner shown in Figure 4.1. We will also briefly describe several useful statistical tests that are popular in the paleomagnetic literature.

In order to calculate mean directions with confidence limits, paleo-magnetists rely heavily on the special statistics known as *Fisher statistics* (Fisher [1953]), which were developed for assessing dispersion of unit vectors on a sphere. Paleomagnetic directional data are subject to a number of factors that lead to scatter. These include:

1) uncertainty in the measurement caused by instrument noise or sample alignment errors,
2) uncertainties in sample orientation,
3) uncertainty in the orientation of the sampled rock unit,
4) variations among samples in the degree of removal of a secondary component,
5) uncertainty caused by the process of magnetization, and
6) secular variation of the Earth's magnetic field.

Some of these sources of scatter (e.g., items 1, 2 and perhaps 6 above) lead to a symmetric distribution about a mean direction. Such a distribution is called a *Fisher distribution*, and is the spherical analog of the normal distribution. Other sources contribute to distinctly non-Fisherian scatter. For example, in the extreme case, item four leads to a girdle distribution whereby directions are smeared along a great circle. In most instances, paleomagnetists assume a Fisher distribution for their data because the statistical treatment allows calculation of confidence intervals, comparison of mean directions, comparison of scatter, etc.

We mention briefly an alternative distribution: the *Bingham Distribution* (Bingham [1964]). The Bingham distribution handles bi-modal, elliptically distributed data. It has some advantages over the Fisher approach which requires uni-modal, spherically symmetric data. However, statistical tests such as tests for randomness, uniqueness, etc., that lend paleomagnetism so much power are difficult with Bingham statistics. This book describes an alternative approach that has the flexibility of Bingham statistics but also allows for statistical testing: the bootstrap. The bootstrap will be discussed at length later in this chapter.

4.1. Parameter estimation

The Fisher probability density function (Fisher [1953]) is given by:

$$F = \frac{\kappa}{4\pi \sinh \kappa} \exp{(\kappa \cos \alpha)}, \qquad (4.1)$$

where α is the angle between the unit vector and the true direction and κ is a precision parameter such that as $\kappa \to \infty$, dispersion goes to zero.

Because the intensity of the magnetization has little to do with the validity of the measurement (except for very weak magnetizations), it is customary to assign unit length to all directions. The mean direction is calculated by first converting the individual directions (D_i, I_i) to cartesian coordinates by the methods given in Chapter 1. The length of the resultant vector, R, is given by:

$$R^2 = (\sum_i x_{1i})^2 + (\sum_i x_{2i})^2 + (\sum_i x_{3i})^2, \qquad (4.2)$$

and the cartesian coordinates of the mean direction are given by:

$$\bar{x}_1 = \frac{1}{R}(\sum_i x_{1i}); \quad \bar{x}_2 = \frac{1}{R}(\sum_i x_{2i}); \quad \bar{x}_3 = \frac{1}{R}(\sum_i x_{3i}). \qquad (4.3)$$

The cartesian coordinates can, of course, be converted back to geomagnetic elements $(\bar{D}, \bar{I})$ by the familiar method described in Chapter 1.

[See Example 4.1]

The precision parameter for the Fisher distribution, κ, is estimated by $\kappa \simeq k = \frac{N-1}{N-R}$ (where N is the number of data points). Using this estimate of κ, we estimate the circle of 95% confidence ($p = 0.05$) about the mean, α_{95}, by:

$$\alpha_{95} = \cos^{-1}[1 - \frac{N-R}{R}[(\frac{1}{p})^{\frac{1}{(N-1)}} - 1]]. \qquad (4.4)$$

In the classic paleomagnetic literature, α_{95} was further approximated by:

$$\alpha'_{95} \simeq \frac{140}{\sqrt{kN}},$$

which is reliable for k larger than about 25 (see Tauxe et al. [1991]).

Another useful parameter is the so-called *circular standard deviation* (CSD, also sometimes called the angular standard deviation), which is approximated by:

$$CSD \simeq \frac{81}{\sqrt{k}}.$$

[See Example 4.2]

If directions are converted to VGPs as outlined in Chapter 1, the transformation distorts a rotationally symmetric set of data into an elliptical distribution. The associated α_{95} may no longer be appropriate and many paleomagnetists (e.g., McElhinny [1973]) calculate the following for 95% confidence regions in VGPs.

$$dm = \alpha_{95} \frac{\cos \lambda}{\cos I}$$

$$dp = \frac{1}{2}\alpha_{95}(1 + 3\sin^2\lambda), \qquad (4.5)$$

where dm is the uncertainty in the paleomeridian (longitude), dp is the uncertainty in the paleoparallel (latitude), and λ is the site paleolatitude.

Two examples of Fisher distributions, one with a large degree of scatter ($\kappa=5$) and one that is relatively tightly clustered ($\kappa=50$) are shown in Figure 4.2. Also shown are the Fisher mean directions and α_{95}s for each data set.

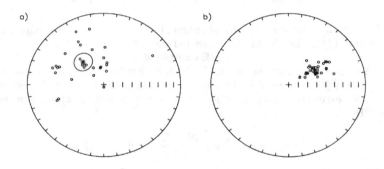

Figure 4.2.   Two Fisher distributions: a) $\kappa = 5$, b) $\kappa = 50$. Mean directions are shown as asterisks, and α_{95}s as ellipses.

The Fisher distribution allows a number of questions to be asked about paleomagnetic data sets.

1) Is a given data set random?
2) Is the mean direction of a given (Fisherian) data set different from some known direction?

3) Are two (Fisherian) data sets different from each other?
4) If a given site has some samples that allow only the calculation of a best-fit plane and not a directed line, what is the site mean direction that combines the best-fit lines and planes?

In the following discussion, we will briefly summarize ways of addressing these issues using Fisher techniques.

4.2. Watson's test for randomness

Watson [1956] demonstrated how to test a given directional data set for complete randomness. His test relies on the calculation of R given by equation 4.2. Because R is the length of the resultant vector, randomly directed vectors will have small values of R, while, for less scattered directions, R will approach N. Watson [1956] defined a parameter R_o that can be used for testing the randomness of a given data set. If the value of R exceeds R_o, the null hypothesis of total randomness can be rejected at a specified level of confidence. If R is less than R_o, randomness cannot be disproved. He calculated the value of R_o for a range of N for the 95% and 99% confidence levels. Watson [1956] also showed how to estimate R_o by:

$$R_o = \sqrt{7.815 \cdot N/3}. \tag{4.6}$$

We plot R_o versus N in Figure 4.3 using both the exact method (dots) and the estimation given by equation 4.6. The estimation works well for $N > 10$, but is somewhat biased for smaller data sets. The critical values of R for $5 < N < 20$ from Watson [1956] are listed for convenience in Table 4.1.

TABLE 4.1. Critical values of R_o for a random distribution (Watson [1956]).

N	95%	99%	N	95%	99%
5	3.50	4.02	13	5.75	6.84
6	3.85	4.48	14	5.98	7.11
7	4.18	4.89	15	6.19	7.36
8	4.48	5.26	16	6.40	7.60
9	4.76	5.61	17	6.60	7.84
10	5.03	5.94	18	6.79	8.08
11	5.29	6.25	19	6.98	8.33
12	5.52	6.55	20	7.17	8.55

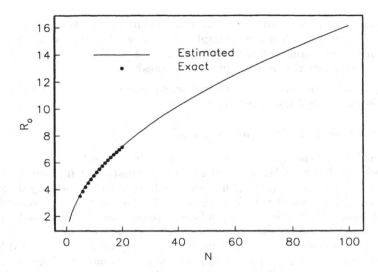

Figure 4.3. Values of R_o calculated by equation 4.6 (line) and exactly (dots) for 95% level of confidence. Exact data are from Watson [1956].

The test for randomness is particularly useful for determining if, for example, the directions from a given site are randomly oriented (the data for the site should therefore be thrown out). Also, one can determine if directions from conglomerate clasts are randomly oriented in the conglomerate test (see Chapter 3).

4.3. Comparing known and estimated directions

The calculation of confidence regions for paleomagnetic data is largely motivated by a need to compare estimated directions with either a known direction (for example, the present field) or another estimated direction (for example, that expected for a particular paleopole). Comparison of a paleomagnetic data set with a given direction is straightforward using Fisher statistics. If the known test direction lies outside the confidence interval computed for the estimated direction, then the estimated and known directions are different at the specified confidence level. The case in which two Fisher distributions are compared can also be simple. If the two confidence regions do not overlap, the two directions are different at the given level of certainty. When one confidence region includes the mean of the other set of directions, the difference in directions is not significant. But, when the two confidence regions overlap, and neither includes the mean of the other, determining the significance of the difference becomes more difficult. For special cases where the two data sets are Fisher distributed with the same dispersion, criteria have been devleloped to test the significance of the difference in data sets by comparing R values for the two data sets separately and together. The Watson test calculates the statistic F as follows:

$$F = (N - 2)\frac{(R_1 + R_2 - R)}{(N - R_1 - R_2)}, \tag{4.7}$$

where R_1, R_2, and R are the resultants of the first, second, and combined data sets, respectively, and N is the total number of data points in both data sets. If F exceeds the value given in an F table for 2 and $2(N\text{-}2)$ degrees of freedom, the null hypothesis that the two data sets have a common mean can be rejected. The Watson approach was extended to include data sets with different degrees of scatter by McFadden and Lowes [1981]. We will discuss an alternative approach for testing for common mean later in the chapter.

4.4. Combining vectors and great circles

Consider the demagnetization data shown in Figure 4.4 for various samples from a certain site. The data from sample *tst1a* seem to hover around a given direction; a mean direction for the last few demagnetization steps can be calculated using Fisher statistics. We can calculate a best-fit line from the data for sample *tst1b* (Figure 4.4b) using the principal component method of Kirschvink [1980] as outlined in Chapter 3. The data from *tst1c* track along a great circle path and can be used to calculate the pole to the best-fit plane. McFadden and McElhinny [1988] outlined a method for estimating the mean direction and the α_{95} from sites that mix planes (great circles on an equal area projection) and directed lines.

- Calculate M directed lines and N great circles using principal component analysis or Fisher statistics.
- Assume that the primary direction of magnetization for the samples with great circles lies somewhere along the great circle path (i.e., within the plane).
- Assume that the set of M directed lines and N unknown directions are Fisher distributed.
- Iteratively search along the great circle paths for directions that maximize the resultant vector R for all the $M + N$ directions.
- Having found the set of N directions that lie along their respective great circles, estimate the mean direction using equation 4.3 and κ as:

$$k = \frac{2M + N - 2}{2(M + N - R)},$$

The cone of 95% confidence about the mean is given by:

$$\cos \alpha_{95} = 1 - \frac{N' - 1}{kR}, [(\frac{1}{p})^{1/(N'-1)} - 1],$$

where $N' = M + N/2$ and $p = 20$.

[See Example 4.3]

4.5. Inclination only data

A different problem arises when only the inclination data are available as in the case of unoriented drill cores. Cores can be drilled and arrive at the surface in short, unoriented pieces. Specimens taken from such core material will be oriented with respect to the vertical, but the declination data are unknown. It is often desirable to estimate the true Fisher inclination of data set having only inclination data, but how to do this is not obvious. Consider the data in Figure 4.5. The true Fisher mean declination and inclination are shown by the asterisk. If we had only the inclination data and calculated a gaussian mean ($< I >$), the estimate would be too shallow as pointed out earlier.

Several investigators have addressed the issue of inclination-only data (Briden and Ward [1966], Kono [1980], McFadden and Reid [1982]). The approach of Briden and Ward [1966] was graphical and not suited for computational ease. McFadden and Reid [1982] developed a maximum likelihood estimate for the true inclination which is less biased than that of Kono [1980] and can be written into a computer program. We outline their approach in the following.

We wish to estimate the co-inclination ($\alpha = 90 - I$) of N Fisher distributed data (α_i), the declinations of which are unknown. We define the estimated value of α to be $\hat{\alpha}$. McFadden and Reid showed that $\hat{\alpha}$ is the solution of:

$$N \cos \hat{\alpha} + (\sin^2 \hat{\alpha}) \sum \cos \alpha_i - 2 \sin \hat{\alpha} \cos \hat{\alpha} \sum \alpha_i = 0,$$

which can be solved for numerically.

They further define two parameters S and C as:

$$S = \sum \sin (\hat{\alpha} - \alpha_i),$$
$$C = \sum \cos (\hat{\alpha} - \alpha_i).$$

An unbiassed approximation for the Fisher parameter κ, k is given by:

$$k = \frac{N - 1}{2(N - C)}.$$

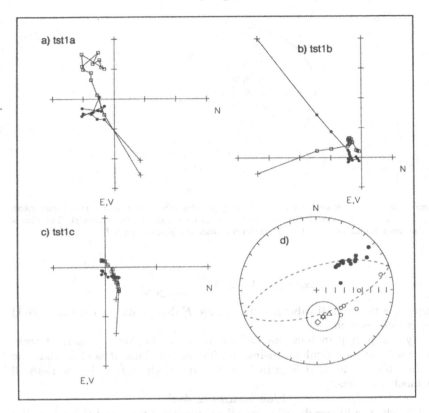

Figure 4.4. Examples of demagnetization data from a site whose mean is partially constrained by a great circle (see Example 4.3). The samples a)*tst1a*, b)*tst1b* and c) *tst1c* which are sibling samples from the same reversely magnetized site. The demagnetization data are plotted as orthogonal projections. The directional data from *tst1c* do not define a single component, but describe a great circle as shown in d). The sample *tst1b* allows calculation of a principal component whose direction is plotted as a diamond in d). Specimen *tst1a* has data that do not converge to the origin. A mean direction was calculated for this sample by standard Fisher statistics and is plotted as a triangle in d. The best-fit great circle and two directed lines allow a mean (star) and associated α_{95} to be calculated using the method of McFadden and McElhinny [1988].

The unbiased estimate $\hat{I}$ of the true inclination is:

$$\hat{I} = 90 + \hat{\alpha} + \frac{S}{C}.$$

Finally, the α_{95} is estimated by:

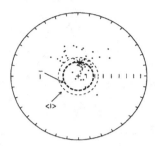

Figure 4.5. Directions drawn from a Fisher distribution with a near vertical true mean direction. The Fisher mean direction from the sample is shown by a asterisk. The Gaussian average inclination $(< I >)$ is shallower than the Fisher mean $\bar{I}$.

$$\cos \alpha_{95} = 1 - 1/2(\frac{S}{C})^2 - \frac{f}{2Ck},$$

where f is the critical value taken from the F distribution with 1 and (N-1) degrees of freedom.

By comparing inclinations estimated using the McFadden-Reid technique with those calculated using the full vector data, it is clear that the method breaks down at high inclinations and high scatter, but works well for most data sets.

[See Example 4.4]

Clearly, the Fisher distribution allows powerful tests and this power lies behind the popularity of paleomagnetism in solving geologic problems. The problem is that these tests require that the data be Fisher distributed. How can we tell if a particular data set is Fisher distributed? What do we do if the data are not Fisher distributed? These questions are addressed in the rest of the chapter.

4.6. Is a data set Fisher distributed?

Let us now consider how to determine whether a given data set is Fisher distributed. The first step is to calculate the orientation matrix **T** of the data and the associated eigenvectors V_i and eigenvalues τ_i (Chapter 3). Substituting the direction cosines relating the geographic coordinate system **X** to the coordinate system defined by **V**, the eigenvectors, where **X** is the "old" and **V** is the "new" set of axes in equation 3.1, we can transform the coordinate system for a set of data from "geographic" coordinates (Figure 4.6a) where the vertical axis is the center of the diagram, to the "data" coordinate system, (Figure 4.6b). where the principal eigenvector ($\mathbf{V}_1$) lies at the center of the diagram, after transformation into "data" coordinates.

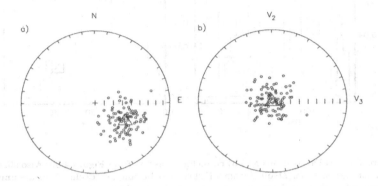

Figure 4.6. Transformation of coordinates from a) geographic to b) "data" coordinates. The direction of the principal eigenvector $\mathbf{V}_1$ is shown by the triangle in both plots.

Equation 4.1 for the Fisher distribution function suggests that declinations are randomly distributed about the mean. In "data" coordinates, this means that the declinations are uniformly distributed from $0 \to 360°$. Furthermore, the probability P of finding a direction within the band:

$$P = F \sin \alpha = \frac{\kappa}{4\pi \sinh \kappa} \exp\left(\kappa \cos \alpha\right) \sin \alpha. \qquad (4.8)$$

Let us compare the data from Figure 4.6 to the expected distributions for a Fisher distribution with $\kappa = 20$ (Figure 4.7). Because the data were generated using the method outlined by Fisher et al. [1987], that draws directions from a Fisher distribution with a specified κ. We used a κ of 20, and it should come as no surprise that the data fit the expected distribution rather well. But how well is "well" and how can we tell when a data set *fails* to be fit by a Fisher distribution? Fisher et al. [1987] describe several methods for doing this, and we outline one of these here.

We wish to test whether the declinations are uniformly distributed and whether the inclinations are exponentially distributed as required by the Fisher distribution. Plots such as those shown in Figure 4.7 are not as helpful for this purpose as a plot known as a *Quantile-Quantile* (Q-Q) plot (see Fisher et al. [1987]). In a Q-Q plot, the data are graphed against the

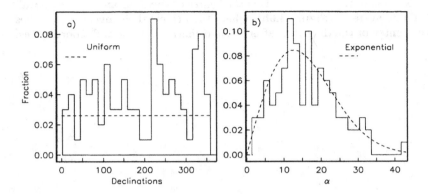

Figure 4.7. a) Declinations and b) co-inclinations (α) from Figure 4.6b. Also shown are behaviors expected for D and I from a Fisher distribution, i.e., declinations are uniformly distributed while co-inclinations are exponentially distributed.

value expected from a particular distribution. Data compatible with the chosen distribution plot along a line. In order to do this, we proceed as follows (Figure 4.8):

• Sort the variable of interest ζ_i into ascending order so that ζ_1 is the smallest and ζ_N is the largest.

• If the data are represented by the underlying density function as in Figure 4.8a, then the ζ_i's divide the curve into $(N+1)$ areas, A_i, the average value of which is $a = 1/(N+1)$. If we assume a form for the density function of ζ_i, we can calculate numbers z_i, that divide the theoretical distribution into areas a_i each having an area a (see Figure 4.8b).

• An approximate test for whether the data ζ_i are fit by a given distribution is to plot the pairs of points (ζ_i, z_i), as shown in Figure 4.8c. If the assumed distribution is appropriate, the data will plot as a straight line.

• The density function P is the distribution function F times the area, as mentioned before. The z_i are calculated as follows:

$$F(z_i) = (i - \tfrac{1}{2})/n, \text{where} i = 1, \ldots, n, \qquad (4.9)$$

so that:

$$z_i = F^{-1}((i - \tfrac{1}{2})/n), \text{where} i = 1, \ldots, n, \qquad (4.10)$$

and where F^{-1} is the inverse function to F. If the data are uniformly distributed (and constrained to lie between 0 and 1), then both $F(x)$

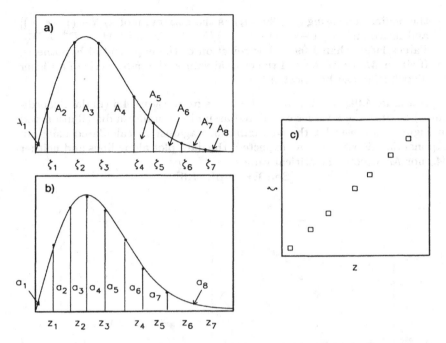

Figure 4.8. a) Illustration of how the sorted data ζ_i divide the density curve into areas A_i with an average area of $1/(N+1)$. b) The values of z_i which divide the density function into equal areas $a_i = 1/(N+1)$. c) Q-Q plot of z and ζ.

and $F^{-1}(x) = x$. For an exponential distribution $F(x) = 1 - e^{-x}$ and $F^{-1}(x) = -\ln(1-x)$.

• Finally, we can calculate parameters M_u and M_e which, when compared to critical values, allow rejection of the hypotheses of uniform and exponential distributions, respectively. To do this, we first calculate:

$$D_N^+ = \text{maximum}[\frac{i}{N} - F(x)], \qquad (4.11)$$

and

$$D_N^- = \text{maximum}[F(x) - \frac{(i-1)}{N}]. \qquad (4.12)$$

For a uniform distribution $F(x) = x$, so M_u is calculated by first calculating D_N^+ as the maximum of $[i/N - \zeta_i]$ and D_N^- as the maximum of $[\zeta_i - (i-1)/N]$. M_u is $D_N^+ + D_N^-$. A value of $M_u > 1.207$ (see Fisher et al [1987]) can be grounds for rejecting the hypothesis of uniformity at the 95% level of certainty. Similarly, D_N^+ and D_N^- can be calculated for the inclination (using $\zeta_i = 90 - I_i$) as the maximum of $[i/N - (1 - e^{-\zeta_i})]$ and maximum of $[(1 - e^{-\zeta_i}) - (i - 1)/N]$ respectively. $M_e = D_N^+ + D_N^-$. Values larger than 1.094 allow rejection of the exponential hypothesis. If either M_u or M_e exceed the critical values, the hypothesis of a Fisher distribution can be rejected.

In Figure 4.9a, we plot the declinations from Figure 4.6 (in data coordinates) against the z values calculated assuming a uniform distribution and in Figure 4.9b, we plot the co-inclinations against zs calculated using an exponential distribution. As expected, the data plot along lines and neither M_u nor M_e exceed the critical values.

[See Example 4.5]

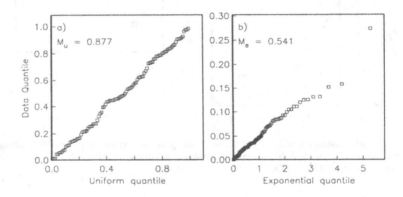

Figure 4.9. a) Quantile-quantile plot of declinations (in data coordinates) from Figure 4.6 plotted against an assumed uniform distribution. b) Same for inclinations plotted against an assumed exponential distribution.

The question arises: what do we do if a particular data set is not adequately fit by a Fisher distribution? The tests designed for Fisher distributions that allow us to ask whether a given data set is different from another, or from an expected, direction will not work reliably unless the data sets are Fisherian. In the following sections we will describe methods for performing similar tests on data that need not be Fisher distributed.

4.7. Non-parametric analysis of vectors

Paleomagnetists have depended since the 1950's on the special statistical framework developed by Fisher [1953] for the analysis of unit vector data.

The power and flexibility of a variety of tools based on Fisher statistics enables quantification of parameters such as the degree of rotation of a crustal block, or whether the geomagnetic field really averages to a geocentric axial dipole independent of polarity. These tools, however, require that the paleomagnetic data belong to a particular parametric distribution. In many geologically important situations, the Fisher distribution fails to model the data adequately. For example, the magnetic field exists in two stable polarity states but the Fisher distribution is uni-modal. The transformation of directions to VGPs tends to distort rotationally symmetric data into elliptical distributions. Furthermore, remanence vectors composed of several components tend to form streaked distributions. Similarly, structural complications (e.g., folding) can lead to streaked distributions of directional data. Thus, non-Fisherian data are a fact of paleomagnetic life. The Fisher-based tests can, at times, be inappropriate and could result in flawed interpretations.

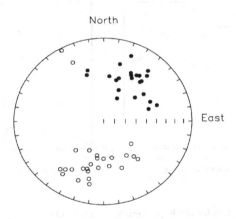

Figure 4.10.   Hypothetical non-Fisherian data set. Normal and reversed polarity data that are not symmetrically distributed. Filled (open) circles plot on the lower (upper) hemisphere.

In the previous sections we learned the basics of Fisher statistics and how to test data sets against a Fisher model. For the rest of the chapter, we will discuss non-parametric methods for performing similar tests that work on non-Fisherian data sets. A non-parametric method is one which does

not require data to be distributed according to some known distribution
that can be characterized with a few parameters.

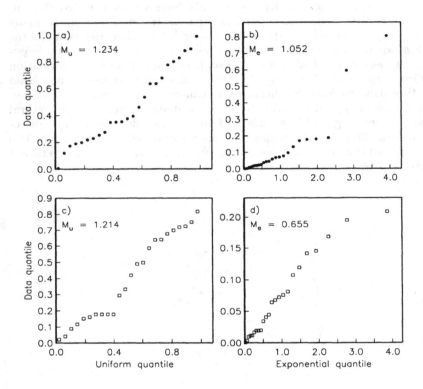

Figure 4.11. Q-Q plots for the hypothetical data set shown in Figure 4.10: a) normal declinations, b) normal co-inclinations, c) reversed declinations, d) reversed co-inclinations.
Neither the normal nor the reversed declinations are well fit by a Fisher distribution.

4.8. Non-Fisherian directional data

The data shown in Figure 4.10 have two polarities and appear to have a
more elliptical distribution than the symmetrical distribution required for
a Fisherian data set. Using the Q-Q method outlined earlier (Figure 4.11),
we find that the data set is probably not rotationally symmetric (M_u is
large) hence it is not well fit by a Fisher distribution.

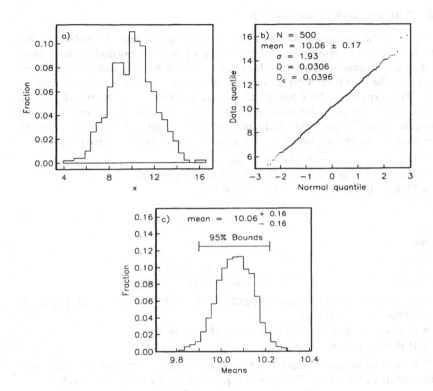

Figure 4.12. Bootstrapping applied to a normal distribution. a) 500 data points are drawn from a Gaussian distribution with mean of 10 and a standard deviation of 2. b) Q-Q plot of data in a). The 95% confidence interval for the mean is given by Gauss statistics as ± 0.17. 10,000 new (para) data sets are generated by randomly drawing N data points from the original data set shown in a). c) A histogram of the means from all the para-data sets. 95% of the means fall within the interval $10.06^{+0.16}_{-0.16}$, hence the bootstrap confidence interval is similar to that calculated with Gaussian statistics.

In order to accomodate data such as those plotted in Figure 4.10, Tauxe et al. [1991] developed an alternative way of characterizing uncertainties for unit vectors. Their method is based on a technique known as the statistical *bootstrap* (see Efron [1982]).

4.9. The statistical bootstrap

In Figure 4.12, we illustrate the essentials of the statistical bootstrap. We will develop the technique using data drawn from a normal distribution. First, we generate a synthetic data set by drawing 500 data points from a normal distribution with a mean $\bar{x}$ of 10 and a standard deviation σ of 2. The synthetic data are plotted as a histogram in Figure 4.12a. In Figure 4.12b we plot the data as a Q-Q plot against the z_i expected for a normal distribution (see Abramowitz and Stegun [1970]). In order to calculate the appropriate values for z_i assuming a normal distribution:

- For $i = 1 \rightarrow N$, calculate $p = \frac{i}{N+1}$.
- If $p > 0.5$, then $q = 1 - p$; if $p < 0.5$, then $q = p$.
- Calculate the following for all $p \neq 0.5$:

$$t = \sqrt{-2\ln^{-1}(q)},$$

and

$$u = t - \frac{(a_1 + a_2 t + a_3 t^2)}{(1 + a_4 t + a_5 t^2 + a_6 t^3)},$$

where $a_1 = 2.515517, a_2 = 0.802853, a_3 = 0.010328, a_4 = 1.432788, a_5 = 0.189269, a_6 = 0.001388$.

- If $p > 0.5$, then $z_i = u$; if $p < 0.5$, then $p = -u$.
- If $p = 0.5$, then $z_i = 0$.

The values of z_i calculated in this way for the simulated Gaussian distribution are plotted as the "normal quantile" data in Figure 4.12b.

[See Examples 4.6 & 4.7]

The simulated data fall along a line in Figure 4.12b which suggests that they are normally distributed (as one would hope). To test this in a more quantitative way, we can calculate D_N^+ and D_N^- as follows:

- Calculate the mean $\bar{x}$ and standard deviation σ for the data.
- Then calculate:

$$p = \frac{x_i - \bar{x}}{\sqrt{2}\sigma},$$

and

$$q = \frac{1}{1 + 0.3275911|x|}.$$

- Substitute q into the following expression (function 7.1.26 from Abramowi and Stegun [1970]):

$$\text{erf}(q) = 1 - e^{-p^2}[a_1 q + a_2 q^2 + a_3 q^3 + a_4 q^4 + a_5 q^5],$$

where $a_1 = 0.254829592, a_2 = -0.284496736, a_3 = 1.421413741$, and $a_5 = 1.061405429$.

• Change the sign of $\text{erf}(q)$ such that it has the same sign as q.

• Substitute $F(x) = 0.5(1 + \text{erf}(q))$ into equations 4.11 and 4.12 for D_N^+ and D_N^- respectively. The Kolmogorov-Smirnov parameter D (e.g., Fisher et al. [1987]) is the larger of D_N^+ or D_N^-.

• The null hypothesis that a given data set is normally distributed can be rejected at the 95% level of confidence if D exceeds a critical value D_c given by $0.886/\sqrt{N}$.

Applying the foregoing to the data in Figure 4.12a yields a D value of 0.0306. Because $N = 500$, the critical value of D, D_c at the 95% confidence level is 0.0396. Happily, our program has produced a set of 500 numbers for which the null hypothesis of a normal distribution has not been rejected. The mean is about 10 and the standard deviation is 1.9. The usual Gaussian statistics allow us to estimate a 95% confidence interval for the mean as $\pm 1.96\sigma/\sqrt{N}$ or ± 0.17.

[See Examples 4.8 & 4.9]

In order to estimate a confidence interval for the mean using the bootstrap, we first randomly draw a list of N data by selecting data points from the original data set using a random number generator (see e.g., Press et al. [1986]). Of course, some data points will be used more than once and others will not be used at all. From this "para-data set" we can calculate a mean. We repeat the procedure of drawing para-data sets and calculating the mean many times (say 10,000 times). The "bootstrapped" means are plotted in Figure 4.12c as a histogram. If these are sorted such that the first mean is the lowest and the last mean is the highest, the 95% confidence interval for the mean is bounded by the means between the 250th and the $9,750^{th}$ mean. The 95% confidence interval calculated for the data in Figure 4.12 by bootstrap is about ± 0.16 which is nearly the same as that calculated the Gaussian way. However, the bootstrap required orders of magnitude more calculations than the Gaussian method, hence it is ill-advised to perform a bootstrap calculation when a parametric one is adequate. Nonetheless, if the data are not Gaussian, the bootstrap provides a means of calculating confidence intervals when there is no other satisfactory way. Furthermore, with a modern computer, the time required to calculate the bootstrap illustrated in Figure 4.12 was virtually imperceptible.

Before we extend the bootstrap to unit vectors, it is important to point out that, with the bootstrap, it is assumed that the underlying distribution is represented by the data, demanding that the data sets are rather large. Moreover, the bootstrap estimates are only asymptotically true, meaning that a large number of bootstrap calculations are required for the confidence intervals to be valid.

What we have described so far is a *simple bootstrap*. There is a way of improving bootstrap estimates for small data sets by using the *parametric bootstrap*. If the data in the previous example had some associated uncertainty, we could use the uncertainty to increase the variability of a small data set and improve the estimate of uncertainty in the mean. Let us say that each data point is a measurement whose uncertainty is Gaussian and its standard deviation is known. Instead of making para-data sets with the original data points, we could randomly select a data point as before, but then calculate a substitute data point drawn from a Gaussian distribution with the mean and standard deviation associated with the original data point. In this way, the para-data sets would reflect more of the underlying distribution than a data set with too few data points.

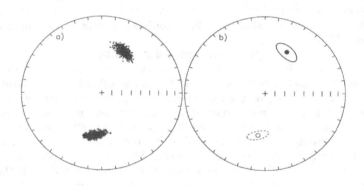

Figure 4.13.   Bootstrap mean directions and 95% confidence ellipses for 500 bootstrap para-data sets drawn from the data shown in Figure 4.10: a) equal area projection of bootstrapped means. Tiny circles (plusses) are lower (upper) hemisphere projections from the normal (reversed) polarity mode of each para-data set. b) The 95% confidence intervals assuming the means from each mode are Kent distributed.

4.10.   A bootstrap for unit vectors

We can now apply the statistical bootstrap to unit vectors:

   • Randomly draw N data points from the data shown in Figure 4.10. This is a para-data set for a simple bootstrap. One can also do a "parametric" bootstrap if each data point represents a site mean of Fisher distributed directions having an estimated κ of k. In this case, draw a

para-data set by random selection of data points and by calculation of
a substitute data point by sampling from a Fisher distribution with the
same N, κ, and mean direction.

• Because paleomagnetic data are often bimodal, the data must be
split into normal and reversed polarity modes. To do this, calculate
the orientation matrix and its eigenparameters (Chapter 3) for each
para-data set. In a Fisher distribution, the Fisher mean coincides with
principal eigenvector $\mathbf{V}_1$, but this is not exactly true with non-Fisherian
data. If the coordinates of the para-data set are transformed to data
coordinates as described in Chapter 3, the transformed inclinations can
be used to assign polarity in an automated and unbiased way; positive
inclinations belong to one mode and negative inclinations to the other.

• Calculate a Fisher mean for each para-data set. Alternatively, if a
more robust estimate of the "average" direction is desired, calculate
the principal eigenvector $\mathbf{V}_1$, which is less sensitive to the presence of
outliers (see, e.g., Tauxe et al. [1991]).

• Repeat the above procedure N_b (say 1,000) times. Examples of a set
of bootstrapped means from the data shown in Figure 4.10 are plotted
in Figure 4.13.

• Now we can estimate the region of 95% confidence for the bootstrapped means or $\mathbf{V}_1$s. A non-parametric approach would be to draw a contour enclosing 95% of the bootstrapped means, or to plot histograms of
cartesian coordinates with 95% confidence limits, as in Figure 4.12c. In
many applications, paleomagnetists desire a more compact way of ex-
pressing confidence regions (for example, to list them in a table) and this
necessitates some parametric assumptions about the distribution of the
means. For this limited purpose, approximate 95% confidence regions
can be estimated by noting the elliptical distribution of the bootstrapped means and by assuming that they are Kent [1982] distributed. The
Kent distribution is the elliptical analogue of the Fisher distribution
(Kent [1982]) and is given as follows:

$$F = c(\kappa, \beta)^{-1} \exp\left(\kappa \cos \alpha + \beta \sin^2\alpha \cos 2\phi\right). \qquad (4.13)$$

where α is the angle between a given direction and the true mean direction (estimated by the principal eigenvector $\mathbf{V}_1$ of the orientation
matrix $\mathbf{T}$), and ϕ is an angle in the plane perpendicular to the true direction with $\phi = 0$ parallel to the major eigenvector $\mathbf{V}_2$ in that plane. κ
is a concentration parameter similar to the Fisher κ, and β is the "oval-
ness" parameter. $c(\kappa, \beta)$ is a complicated function of κ and β (Fisher et
al. [1987]), When β is zero, the Kent distribution reduces to a Fisher
distribution.

• The parameters of interest are calculated by rotating the set of boot-strapped means from a given mode x into the data coordinates x' by the transformation:

$$x' = \mathbf{\Gamma}^T x, \tag{4.14}$$

where $\mathbf{\Gamma} = (\boldsymbol{\gamma}_1, \boldsymbol{\gamma}_2, \boldsymbol{\gamma}_3)$, and the columns of $\mathbf{\Gamma}$ are called the constrained eigenvectors of $\mathbf{T}$. The vector $\boldsymbol{\gamma}_1$ is parallel to the Fisher mean of the data, whereas $\boldsymbol{\gamma}_2$ and $\boldsymbol{\gamma}_3$ (the major and minor axes) diagonalize $\mathbf{T}$ as much as possible subject to being constrained by $\boldsymbol{\gamma}_1$ (see Kent [1982], but note that his x_1 corresponds to x_3 in conventional paleomagnetic notation). The following parameters may then be computed

$$\hat{\mu} = N^{-1}\sum_k x_{k1}'$$
$$\hat{\sigma}_2^2 = N^{-1}\sum_k (x_{k2}')^2 \tag{4.15}$$
$$\hat{\sigma}_3^2 = N^{-1}\sum_k (x_{k3}')^2.$$

As defined here, $\hat{\mu} = R/N$ (R is closely approximated by equation 4.2). Also to good approximation, $\hat{\sigma}_2^2 = \tau_2$, and $\hat{\sigma}_3^2 = \tau_3$, where τ_i are the eigenvalues of the orientation matrix. The semi-angles ζ_{95} and η_{95} subtended by the major and minor axes of the 95% confidence ellipse are given by:

$$\zeta_{95} = \sin^{-1}(\sigma_2\sqrt{g}), \quad \eta_{95} = \sin^{-1}(\sigma_3\sqrt{g}), \tag{4.16}$$

where $g = -2\ln(0.05)/(N\hat{\mu}^2)$.

The tensor $\mathbf{\Gamma}$ is, to a good approximation, equivalent to $\mathbf{V}$, the eigenvectors of the orientation matrix. Therefore, the eigenvectors of the orientation matrix $\mathbf{V}$ give a good estimate for the directions of the semi-angles by:

$$D_\zeta = \tan^{-1}(v_{22}/v_{12}), \quad \text{and} \quad I_\zeta = \sin^{-1}v_{32},$$
$$D_\eta = \tan^{-1}(v_{23}/v_{13}), \quad \text{and} \quad I_\eta = \sin^{-1}v_{33}, \tag{4.17}$$

where for example the x_2 component of the smallest eigenvector $\mathbf{V}_3$ is denoted v_{23}. The 95% confidence ellipses calculated in this way for our example data set are shown in Figure 4.13b.

[See Example 4.10]

4.11. The parametric bootstrap

The bootstrap described here is a "simple" or "naive" bootstrap in that no distribution is assumed for the data. We must assume, however, that all the uncertainty inherent in the data is reflected in the data distribution. If the data set is smaller than about $N = 20$, this leads to uncertainty ellipses that are too small (Tauxe et al. [1991]). Many paleomagnetic data sets are

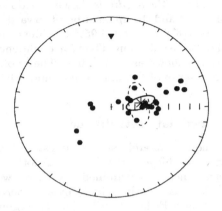

Figure 4.14. Normal polarity data from Figure 4.10 transformed into VGPs and plotted in polar equal area projection. The dashed ellipse is obtained from VGP transformation of α_{95}, while the solid ellipse is determined by a bootstrap.

smaller than this, yet they are demonstrably non-Fisherian. However, if we are able to assume some parametric form for data from e.g., a given site, we can perform a superior technique which is known as the *parametric bootstrap*. As applied here, we assume that each site with N_s samples is Fisher distributed (in principle, a testable assumption). Then, after random selection of a particular site for inclusion in the para-data set, we draw N_s new directions from a Fisher distribution with the same mean direction, κ and N. From these, we then calculate a substitute mean direction, and use that in the para-data set. Otherwise, we follow the same procedure as before.

For large data sets ($N > 25$), the parametric and simple bootstraps yield very similar confidence ellipses. For smaller data sets, the parametric ellipses are larger, and are probably more realistic.

4.12. Application to VGPs

We stated earlier that the transformation of data from directions to VGPs can cause a distortion from rotationally symmetric data to elliptically distributed data. To illustrate what happens if data are distributed as in our hypothetical normal data set, we calculate VGPs in Figure 4.10, the ellipse described by dp, dm, and our bootstrap confidence ellipses. The VGPs are plotted in Figure 4.14 and the dp, dm is shown as a dotted line. Because the declinations are smeared and the dp must point towards the site (shown as a triangle), the long axis of the so-called 95% confidence regions is perpendicular to the actual data distribution. The 95% confidence ellipse, calculated using the bootstrap, is shown as a solid line. The bootstrapped confidence ellipse gives a better sense of the uncertainty and follows the trend in the data.

4.13. When are two data sets distinct?

The test for a common mean addresses the question "can the means of two data sets be discriminated from one another?" Another way of putting it is, "If a set of bootstrap means is examined, are there two distinct groups or is there just one?" We explore these ideas by considering several data sets drawn at random from a Fisher distribution. In Figure 4.15 we show the means of two data sets (D_1 and D_2), each drawn from distributions with a κ of 10. The mean direction of each lies outside the confidence region of the other, but the F test of Watson [1956] gives a value of 2.03, which passes the test for a common mean (see Section 4.3).

In order to compare the two populations of bootstrapped means, we convert them to cartesian coordinates. Histograms of the cartesian coordinates of the bootstrap means (Figure 4.15) are uni-modal, which suggests a common mean. As a more quantitative test, we plot the intervals from each data set that contain 95% of the respective sets of cartesian coordinates of the bootstrapped means. These overlap, which confirms that the two data sets cannot be distinguished at the 95% confidence level.

In Figure 4.16, we show means and α_{95}s from a different pair of Fisher distributions. The α_{95}s still overlap one another, but they fail the Watson test for a common mean ($F = 3.9$). In this case, the histograms of the x_2 cartesian coordinates of the bootstrap means have twin peaks, which is consistent with the presence of two directions. Furthermore, the 95% confidence bounds do not overlap, thereby providing a quantitative bootstrap test for a common mean.

[See Example 4.11]

4.14. Application to the "reversals test"

The so-called *reversals test* in paleomagnetism constitutes a test for a common mean for two modes, one of which has been "flipped" to its antipode. We apply our bootstrap test for common mean to the data shown in Figure 4.10. The histograms of the cartesian coordinates of the bootstrapped means are shown in Figure 4.17. There are two "humps" in the bootstrap test. However, the confidence intervals for the normal and reversed an-

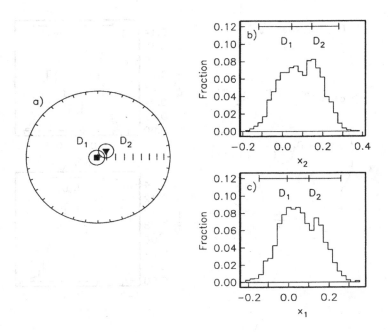

Figure 4.15. The bootstrap test for a common mean. a) Equal-area projections of means and α_{95}s of two simulated Fisherian data sets (D_1 and D_2), each with κ of 10. D_2 (triangle) has a mean that lies outside the confidence interval of D_1 (square). b) Histograms of x_2 components of the bootstrapped means from 500 para-data sets. Also shown are the bounds for each data set that include 95% of the components. The confidence intervals for the different data sets (D_1 and D_2) overlap. c) Same as b) but for the x_1 component. Because the confidence intervals for all components overlap (x_3 not shown), the two sets of bootstrapped means cannot be distinguished at the 95% level of confidence; they pass the bootstrap test for common mean.

tipodes overlap, thereby suggesting that the two means cannot be distinguished at the 95% level of confidence. Thus, the data in Figure 4.10 pass the bootstrap reversals test.

[See Example 4.12]

4.15. The bootstrap fold test

A final test is extremely useful in paleomagnetism: the fold test (Chapter 3). One of the key components in paleomagnetic studies is to determine the

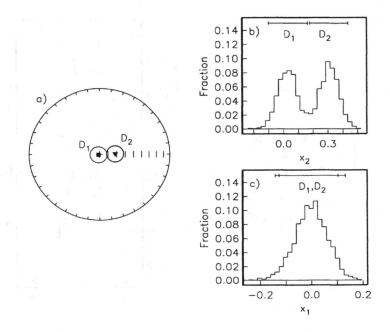

Figure 4.16. Same as Figure 4.15, but the data sets fail both the Watson [1956] and bootstrap tests for common mean.

coordinate system (geographic, tilt adjusted or somewhere in between) for which the directional data are most tightly clustered. If a rock has moved from its original position, was it magnetized in the original, in the present or in some other position? Moreover, is simple rotation about strike an appropriate method to restore the beds to their original positions? In the classic fold test as first proposed by Graham [1949], the directions of magnetization of a deformed rock unit are assumed to be most closely parallel in the orientation in which the magnetization was acquired. Therefore, if a rock has retained an original magnetization through a subsequent folding or tilting event, the magnetic directions will cluster most tightly after they have been rotated back to their original positions.

The fold test appears at first glance to be simple, but it is not (see, e.g., McFadden and Jones [1981], Fisher and Hall [1990], McFadden [1990], Watson and Enkin [1993], Tauxe and Watson [1994]). The primary problem is that paleomagnetic vectors are never perfectly parallel. The scattered

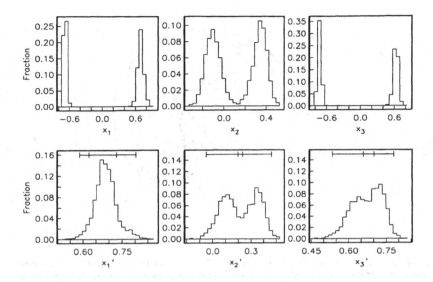

Figure 4.17. Histograms of cartesian coordinates of means of para-data sets drawn from the data shown in Figure 4.10. In the lower plots, the reversed polarity mode has been flipped to the antipode (x_i'). The intervals containing 95% of each set of components are drawn (see Figure 4.15 and 4.16). Because the confidence bounds from the two data sets overlap in all three components, the means of the reversed and normal modes cannot be distinguished at the 95% level of confidence; they pass the bootstrap reversals test.

nature of the data means that a statistical test is necessary to determine whether clustering is "significantly" better in one orientation or another.

Graham [1949] lived in a world with no adequate statistical framework for dealing with directional data, which was first introduced by Fisher [1953]. McElhinny [1964] proposed that the concentration of the data (using Fisher's precision parameter, κ) could be calculated before and after adjusting for bedding tilt and that the ratio of the two values should be compared with those listed in statistical "F" tables. Ratios higher than the "F" value for a given N were deemed to constitute a significant increase in concentration after adjusting for tilt, thus representing a positive fold test. The McElhinny [1964] test can be done on the back of an envelope and was immediately embraced by the paleomagnetic community; it is still in frequent use.

Although its simplicity is a great strength, there are several problems with the McElhinny [1964] fold test (see also McFadden [1990]). First, the

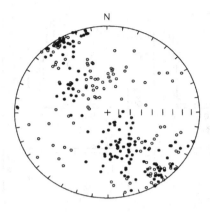

Figure 4.18.   Equal area projection of a set of directions in geographic coordinates. The data are streaked in a girdle distribution and the polarity of many data points is ambiguous. (The data are the "P-Component" of Juárez et al. [1994]; redrawn from Tauxe and Watson [1994].)

geomagnetic field has two preferred states and is not perfectly dipolar. Directions observed in paleomagnetic samples are therefore not only scattered but are often of two polarities. Second, the magnetic directions may be most tightly clustered somewhere other than in "geographic" or 100% tilt adjusted coordinates (see e.g., McCabe et al. [1983]). Finally, structural "corrections" are not perfectly known. Not only are the bedding orientations themselves often difficult to measure accurately, but detection of complications such as plunging folds, and multiple phases of tilting requires extensive field work. It is nearly impossible to assess rotation about the vertical axis on the basis of field relations alone, as it results in no visible effect on the dip of the beds themselves. Because of this uncertainty, we might reasonably ask whether if the data are actually most tightly clustered at, say 90% tilt adjusted (as opposed to 100%), does this constitute a "failed" fold test (see Watson and Enkin [1993]).

We consider first the problem of dual polarity. We plot data from Juárez et al. [1994] in Figure 4.18. These are sample directions, each of which has a tectonic correction. The samples were taken for magnetostratigraphic purposes and one sample constitutes one site. In geographic coordinates (and even in some cases after adjusting for tilt), the polarity is ambiguous and the calculation of κ necessitates using the tilt adjusted data to identify

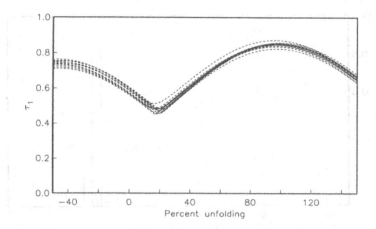

Figure 4.19. The largest eigenvalues τ_1 of the orientation matrices from representative para-data sets drawn from Figure 4.18. The directions are adjusted for tilt incrementally from -50% to 150%. The largest value of τ_1 occurs near 100% in all of the para-data sets.

the polarity of the samples.

[See Example 4.13]

The classic fold test of McElhinny [1964] requires calculation of κ which can only be done with data of a single polarity. Obviously, fold tests that rely on κ will not be straightforward with data such as these. An alternative approach was proposed by Tauxe and Watson [1994], and is based on the orientation matrix. In the orientation matrix, polarity does not play a role and the "tightness" of grouping is reflected in the relative magnitudes of the eigenvalues (τ). As the data become more tightly grouped, the variance along the principal axis grows and those along the other axes shrink. Thus, examination of the behavior of τ_1 during unfolding would reveal the point at which the tightest grouping is achieved, without knowledge of polarity.

Suppose we find that the degree of unfolding required to produce the maximum in τ_1 is 98%. Is this a positive fold test suggesting a pre-folding remanence or is the difference between 98% and 100% significant? For this we call on the familiar bootstrap. Numerous para-data sets can be drawn. We can then calculate the eigenparameters of the orientation matrix for a range of % unfolding. Some examples of the behavior of τ_1 during tilt adjustment of representative para-data sets drawn from the data in Figure 4.18 are shown in Figure 4.19. Figure 4.20 is a histogram of maxima of τ_1 from 500 para-data sets. These are sorted as described in Section 4.9 and

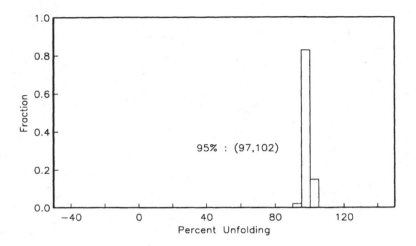

Figure 4.20. Histogram of 500 maxima of τ_1 and the calculated 95% confidence interval. These data "pass" the bootstrap fold test.

the 95% confidence interval for the degree of unfolding required to produce the tightest grouping (the highest τ_1) is thus constrained to lie between 97 and 102%.

The data from Figure 4.18 are shown after 100% tilt adjustment in Figure 4.21. The tilt adjusted data are not only better grouped, but now the polarities of most samples can be readily determined. An advantage of the bootstrap approach is the fact that the data do not need prior editing to split them into normal and reversed polarity groups, which is a particularly onerous task for the data considered here.

For small data sets, such as that of Gilder et al. [1993] (Figure 4.22), we employ a parametric bootstrap, whereby para-data sets are generated by first randomly selecting a site for inclusion, then by drawing a substitute direction from a Fisher distribution having the same D, I, N, and κ. Results from such a parametric bootstrap are shown in Figure 4.22 and both 0 and 100% unfolding are effectively excluded, with the tightest grouping achieved at around 70% unfolding.

It is possible that the magnetic directions shown in Figure 4.22 were acquired during folding; fold test results with the tightest grouping between 0 and 100% unfolding are often interpreted as an indication of "syn-folding remanence" (see McCabe et al. [1983]). However, two phases of folding

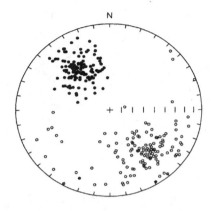

Figure 4.21.   Data from Figure 4.18 after 100% adjusting for tilt. Polarities are more readily identifiable. (Redrawn from Tauxe and Watson, [1994]).

(or one that is not a simple rotation about strike), that are incorrectly "unfolded" can be grouped most tightly between 0 and 100%. To illustrate this, Tauxe and Watson [1994] used a simulated data set drawn from a Fisher distribution having $\kappa = 30$ and a mean direction of North, dipping 45° (shown in Figure 4.23a). The data were then split into two "limbs" and subjected to two rotations (Figure 4.23b). The only observable effect in the field would be the latter, so we performed a "fold test" undoing only the second rotation. The "tilt adjusted" data are shown in Figure 4.23c at both peak concentration (61%) and at 100% unfolding.

The maximum in τ_1 occurs at about 61% (Figure 4.23d), excluding 0 and 100% unfolding as the appropriate coordinate systems at the 95% level of confidence. However, 61% unfolding does not yield precisely the initial distribution because of failure to account for the vertical axis rotation. This simulation produces results similar to those in Figure 4.22 and proves that a peak in concentration between 0 and 100% unfolding does not necessarily imply a syn-folding remanence. Nonetheless, it does indicate that either because of incorrect structural information or because of complications in the remanence, the "100% adjusted" data are not valid for paleomagnetic purposes and justifiably fail the fold test. Of course, it would be dangerous to use the mean direction from even the partially tilt adjusted data for paleomagnetic purposes.

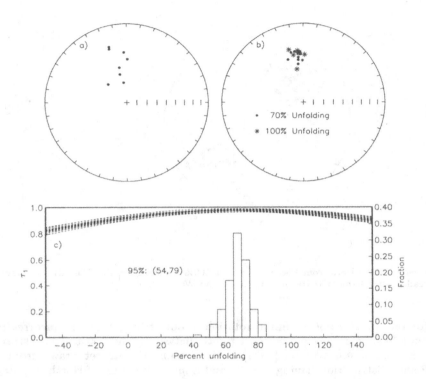

Figure 4.22. Directional data of Gilder et al. [1993] from the Xinlong Formation. The data are site means of characteristic components. The data are shown in a) geographic coordinates and b) after tilt adjustment (70% and 100%). c) τ_1s from 20 para-data sets and a histogram of the 500 maxima in τ_1. Both geographic and 100% tilt adjustment are excluded at the 95% confidence level, thereby suggesting either a complex magnetization, complex folding history or a syn-folding remanence acquisition. (Redrawn from Tauxe and Watson, [1994]).

4.16. Summary

In this chapter, we have reviewed the statistical analysis of unit vectors, as developed by Fisher [1953]. We have also introduced the bootstrap for unit vectors. Bootstrap procedures allow most of the useful statistical tests that are frequently performed on paleomagnetic data to be applied to non-Fisherian data. Many data sets are not Fisher distributed and the use of parametric tests can lead to erroneous conclusions. In contrast, use of bootstrap tests, when the data are Fisherian gives the same result – it just

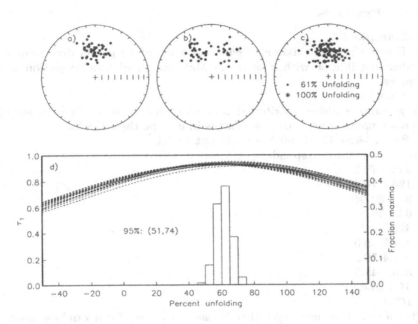

Figure 4.23. Synthetic data illustrating the effect of unobserved vertical axis rotation coupled with rotation about a horizontal axis. a) Data in 'initial' coordinates are drawn from a Fisher distribution with $\kappa = 30$. b) The data were split into two groups (shown as dots and asterisks) and each subjected to two rotations. The dots were rotated first by 20° about a vertical axis, then 45° to the east about a North/South horizontal axis. The asterisks were rotated first by -20° about a vertical axis, then 45° to the west about the horizontal axis. c) The tilt adjustment only accounts for the latter (observable) rotation and d) the fold test was performed as described in the test. Both 'geographic' and 100% tilt adjusted coordinate systems are excluded at the 95% level of confidence. Thus use of partially adjusted data, despite the fact that it represents the tightest concentration of data, is ill-advised. Redrawn from Tauxe and Watson [1994].

takes about 1,000 times longer to get it!

4.17. Examples

• Example 4.1

1) Use the program **fisher** to generate a set of 10 points drawn from a Fisher distribution with $\kappa = 15$. 2) Make an equal area projection with **eqarea**.

Solution

To generate a Fisher distributed data set with $N = 10$ and $\kappa = 15$, save it to a file named **ex4.1** and see what is in it, type the following:

% **fisher -kns 15 10 66** > **ex4.1; cat ex4.1**

and the computer responds:

```
248.0 75.2
196.1 70.4
81.4 81.0
64.1 70.5
33.6 70.2
311.7 80.5
207.4 59.2
318.9 48.3
34.0 60.4
169.1 61.5
```

The **-s** switch is an integer that is used as a seed for a random number generator. Different distributions can be made with different values of **-s**. To make a postscript file of an equal area projection of these data and to view it on the screen, type:

% **fisher -kns 15 10 66** | **eqarea** | **plotxy; ghostview mypost**

This causes **plotxy** to make a postscript file named **mypost** that is shown in Figure 4.24.

• Example 4.2

Calculate a mean direction, k, and α_{95} from the distribution generated in Example 4.1 using the program **gofish**. Repeat for the principal direction using **goprinc**.

Solution

Type the following:

% **gofish** < **ex4.1**

and the computer responds:

```
318.1 88.9 10 9.1220 10.3 15.8
```

To find out what these numbers are, type:

% **gofish -h**

or check Appendix 1.

As a short-cut, you could take advantage of UNIX's pipe facility by typing:

% **fisher -kns 15 10 66** | **gofish**

North

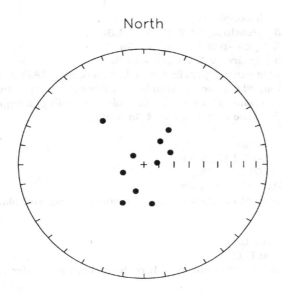

Figure 4.24. Equal area projection of data generated in Example 4.1, using the command: **fisher -kns 15 10 66 | eqarea | plotxy**.

which does the same thing, without the intermediate file **ex4.1**.
Now type:
% **goprinc** < **ex4.1**
to which the computer reponds:
    333.9 89.4 10 0.837
which are the principal directions (V_1) D, I, N and τ_1 respectively.

• **Example 4.3**
1) Use the program **fishdmag** to calculate a mean direction from the last
seven data points of sample *tst1a* in file **ex4.3** and Figure 4.4. 2) Use the
program **pca** to calculate the principal component direction from the last
13 data points of sample *tst1b*. 3) Use program **gtcirc** to calculate the best-
fitting great circle from the last 22 data points of sample *tst1c*. 4) Finally,
use the program **lnp** to calculate the mean direction and α_{95} using the two
directed lines and the great-circle data for site *tst1*.

Solution

To extract data for each sample, type:

```
% grep tst1a ex4.3 | fishdmag -f 7 13 > ex4.3a
% grep tst1b ex4.3 | pca -p 6 17 >> ex4.3a
% grep tst1c ex4.3 | gtcirc -g 2 23 >> ex4.3a
```

The columns are: sample name, f/p/g, first step, last step, α_{95}/MAD/MAD$_{plane}$, declination, inclination, where the "f" stands for Fisher mean, "p" stands for principle component, and "g" stands for the pole to best-fit plane (great circle). To see what is in the output file **ex4.3a**, type:

cat ex4.3a

and the computer responds:

```
tst1a f 7 350.00 550.00 9.3 152.6 -60.7
tst1b p 12 20.00 450.00 22.1 178.1 -54.6
tst1c g 22 1.00 450.00 11.4 341.1 -26.2
```

To calculate an estimate of the Fisher mean of combined lines and planes, type:

% lnp < ex4.3a

and the computer responds:

```
tst1 2 1 76.3 16.7 166.7 -60.1
```

To find out what these numbers are, type **lnp -h** or check Appendix 1.

● **Example 4.4**

Use program **fishrot** to draw a set of 50 data points from a Fisher distribution with a mean declination of 0, inclination of 20, and a κ of 30. Calculate the Fisher statistics of the data set with the program **gofish**. Now **awk** out the inclinations and estimate the mean and 95% confidence bounds using the program **incfish**. Repeat this for populations with a mean inclination of 40, 60 and 80. How well does the inclination-only method work for high inclinations?

Solution

To learn about **fishrot** type:

% fishrot -h

or check Appendix 1.

The following command will generate the desired distribution:

% fishrot -kndi 30 50 0 20 > ex4.4

Calculate the Fisher statistics using:

% gofish < ex4.4

and the computer responds:

```
359.7 19.3 50 48.2450 27.9 3.9
```

Now type the following to select the inclination data (second column) and pipe them to **incfish**:

% awk '{print $2}' ex4.4 | incfish

and the computer responds:

21.0 24.6 17.3 50 27. 3.

To find out what this means, type **incfish -h**, or check Appendix 1.

Note that the estimated mean inclination using **incfish** is 21.4 and that calculated using **gofish** is 21.3. To see what happens at higher inclinations, we can streamline things a bit using the **tee** utility. The **tee** utility transcribes the standard input to the standard output and copies it to a specified file name, as follows:

% **fishrot -kndi 30 50 0 40** | **tee ex4.4a** | **gofish**

to which the computer responds:

359.6 39.3 50 48.2439 27.9 3.9

and silently makes **ex4.4a** containing the **fishrot** output. Now type:

% **awk '{print $2}' ex4.4a** | **incfish**

and get:

40.5 44.3 36.8 50 27. 3.8

also in good agreement with the **gofish** calculation.

Repeat for an inclination of 60:

% **fishrot -kndi 30 50 0 60** | **tee ex4.4b** | **gofish** to which the computer responds:

4.5 61.3 50 48.3150 29.1 3.8

and:

% **awk '{print $2}' ex4.4b** | **incfish**

to get:

59.6 63.8 55.5 50 27. 4.1

Finally, try an inclination of 80:

% **fishrot -kndi 30 50 0 80** | **tee ex4.4c** | **gofish** to which the computer responds:

14.2 81.1 50 48.3154 29.1 3.8

and:

% **awk '{print $2}' ex4.4c** | **incfish**

to get the message:

************Result no good***************

This illustrates the point that the method breaks down at high inclinations.

● **Example 4.5**

Use program **fishqq** to check if a data set produced by **fisher** is likely to be Fisherian.

Solution

Type the following to draw 25 directions from a Fisher distribution having a κ of 25:

% **fisher -kns 25 25 44 > ex4.5**

(To draw a different distribution, you can change the random seed from 44 to any non-zero integer.)

Try plotting the data in an equal area projection with the command:

% eqarea < ex4.5 | plotxy; ghostview mypost
(see Figure 4.25).

North

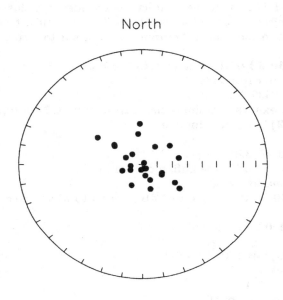

Figure 4.25. Equal area projection from Example 4.5, using the command: **fisher -kns 25 25 55 | eqarea | plotxy**.

To make a Q-Q plot against a Fisher distribution, type:
% fishqq < ex4.5 | plotxy; ghostview mypost
This produces the Q-Q plot shown in Figure 4.26.

• Example 4.6
Use the program **gauss** to generate a data set by drawing 200 data points from a normal distribution with a mean of 22, a standard deviation of 5. Plot these data as a histogram with **histplot** and calculate a mean and standard deviation using **stats**.
Solution
Type the following:
% gauss -msni 22 5 200 22 > ex4.6
to generate a file with the normal polarity data in it. Now type:

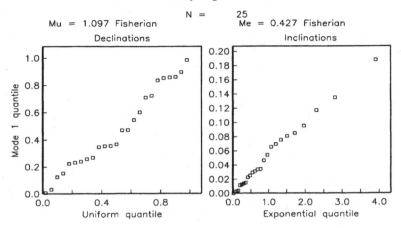

Figure 4.26. Q-Q plot generated in Example 4.5, using the command: **fisher -kns 25 25 44 | fishqq | plotxy**.

% **histplot** < **ex4.6 | plotxy**
to cause plotxy to generate the postscript file **mypost** which can be viewed or plotted as usual (see Figure 4.27).
Finally, type:
% **stats** < **ex4.6**
to get:
    200 22.0720 4414.41 4.89825 22.1921 0.346358 0.678863
To find out what these numbers are, type **stats -h** or check Appendix 1.

• **Example 4.7**
Use the program **qqplot** to plot a Q-Q plot of the data in **ex4.6**, and to calculate the Kolomogorov-Smirnov statistic D, and the 95% asymptotic significance level for N data points.
Solution
Type:
% **qqplot** < **ex4.6 | plotxy**
and view the postscript file **mypost** as in Figure 4.28. D is below the critical value $D_c = 0.886/\sqrt{N}$, so the null hypothesis that the data set is Gaussian cannot be rejected at the 95% level of confidence.

• **Example 4.8**
Use the program **bootstrap** to calculate a bootstrap confidence interval for the data generated in Example 4.6.
Solution

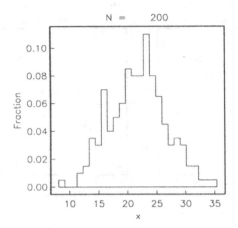

Figure 4.27. Histogram from Example 4.6, using the command: **gauss -msni 22 5 200 22 | histplot | plotxy**.

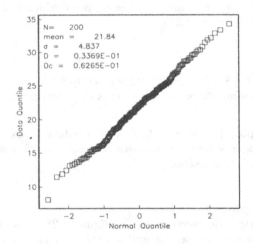

Figure 4.28. Q-Q plot from Example 4.7, using the command: **gauss -msni 22 5 200 22 | qqplot | plotxy**.

Type the following:

% **bootstrap -p** < **ex4.6** | **plotxy**

to generate a distribution of 1,000 bootstrapped means (this is the default number; a different number of bootstraps can be selected using the **-b** switch). The program outputs plotxy commands for the postscript file **mypost** shown in Figure 4.29.

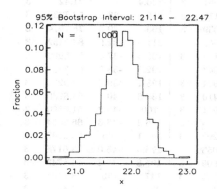

Figure 4.29. Histogram from Example 4.8, using the command: **gauss -msni 22 5 200 22** | **bootstrap -p** | **plotxy**.

• Example 4.9

Use program **bootdi** to see: 1) if the data set in Figure 4.9 (data are in Table 4.2) are likely to be Fisherian; and 2) what the approximate 95% confidence ellipses are (try both simple and parametric bootstraps); and 3) what the approximate 95% confidence ellipses are using the principal eigenvectors instead of Fisher means.

Solution

First enter the data from Table 4.3 into file **ex4.9**. Then type:

% **bootdi** < **ex4.9**

to which the computer responds:

Total N = 49

Mode:	Dec,	Inc,	a95,	N,	kappa,	Fisherian ?
1st:	26.5	39.1	10.7	25	8.	no
2nd:	189.2	-44.3	8.3	24	14.	no
Mode	eta,	dec,	inc,	zeta,	dec,	inc
1st:	6.61	239.44	45.93	11.79	130.96	17.06
2nd:	4.59	152.36	39.38	10.64	259.14	19.38

TABLE 4.2. Data for Example 4.9.

D	I	κ	N	D	I	κ	N
330.9	-3.0	70.1305	3	216.2	-22.8	31.2651	3
333.7	-22.4	69.1498	3	192.2	-23.8	58.4369	3
21.5	16.8	52.5162	3	202.2	-24.5	53.2182	3
39.6	22.7	40.7081	3	220.4	-25.7	21.4247	3
42	27.9	61.1116	3	216.8	-28.8	53.7732	3
15	29.8	60.2039	3	193.7	-29.6	75.7974	3
36.5	32.7	71.5562	3	211.9	-32.5	84.3452	3
33.9	34.2	67.4151	3	196.6	-37.5	55.8909	3
27.8	34.9	32.3343	3	202.3	-38.5	69.2351	3
41.8	36.1	43.4094	3	198.1	-39.7	62.0232	3
343.3	36.5	35.6410	3	139.6	-40.2	61.6730	3
29.2	36.9	45.6071	3	160	-40.2	51.0890	3
72.3	37.0	57.3765	3	186.3	-43.3	66.2564	3
41.1	37.9	28.3202	3	141.1	-45	60.5742	3
31.5	39.5	42.9056	3	173.3	-45.0	38.6178	3
340.8	41.1	56.4909	3	195.8	-45.4	64.5743	3
66.1	42.1	28.8305	3	147.9	-50	48.1485	3
54.4	43.8	43.5158	3	222.5	-51.3	46.4794	3
6.7	44.9	53.8938	3	180.2	-53	71.5305	3
18.3	45.1	40.7290	3	189.4	-53	22.9451	3
73.8	46.2	49.6875	3	168.1	-53.2	28.0272	3
16	47.8	53.3307	3	190.5	-55.8	48.6318	3
51.1	52.7	13.5338	3	131.2	-56.3	30.0795	3
24.5	56.5	84.4026	3	205	-58.5	42.7021	3
28.9	64.5	47.5771	3				

For a parametric bootstrap, type:
% **bootdi -p** < **ex4.9**
to which the computer responds:

Total N = 49

Mode:	Dec,	Inc,	a95,	N,	kappa,	Fisherian ?
1st:	26.5	39.1	10.7	25	8.	no
2nd:	189.2	-44.3	8.3	24	14.	no
Mode	eta,	dec,	inc,	zeta,	dec,	inc
1st:	6.71	240.24	45.67	11.54	131.39	17.52
2nd:	5.30	152.44	39.41	11.25	259.20	19.34

To work on the principal eigenvectors, type:
% **bootdi -P** < **ex4.9**

to which the computer responds:

Total N = 49

Mode:	Dec,	Inc,	a95,	N,	kappa,	Fisherian ?
1st:	30.8	40.9	10.7	25	8.	no
2nd:	9.8	44.1	8.3	24	14.	no
Mode	eta,	dec,	inc,	zeta,	dec,	inc
1st:	6.26	222.02	48.52	9.77	125.69	5.57
2nd:	4.70	150.84	38.69	10.99	258.38	20.62

• Example 4.10

Use **plotdi** to make an equal area projection of the data and bootstrap confidence ellipses of the data in **ex4.9**. Make a plot of the bootstrapped eigenvectors.

To plot the data and the bootstrap ellipses, type:

% **plotdi < ex4.9 | plotxy**

See Figure 4.30.

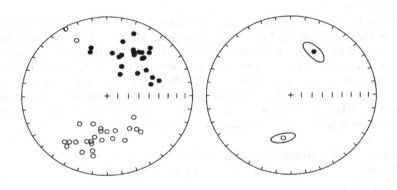

Figure 4.30. Equal area projections from Example 4.10, using the command: **plotdi < ex4.9 | plotxy**.

To plot the data and the bootstrap eigenvectors, type:

% **plotdi -v < ex4.9 | plotxy**

See Figure 4.31.

The command **bootdi -v < ex4.9** will generate the list of bootstrapped eigenvectors used in **plotdi -v**.

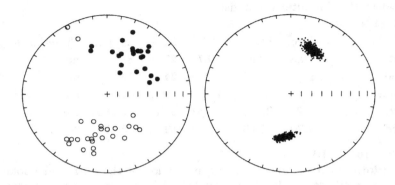

Figure 4.31. Equal area projections from Example 4.10, using the command: **plotdi -v** < **ex4.9**.

• Example 4.11

Assume that the sampling site for the data in Table 4.2 was at latitude λ 35° N and longitude ϕ 120° W. 1) Calculate the direction expected for a dipole field at that location using the dipole formula (tan (I) = 2 tan (λ)) (see Chapter 1). 2) Use **cart_hist** to compare the data set in Table 4.3 with the expected dipole direction. 3) The data were taken in 1997 near sea level. Use **igrf** to calculate the expected geomagnetic field direction. Use **cart_hist** to compare the data with this direction. 4) Use **cart_hist** to perform a parametric bootstrap reversals test.

Solution

1) The expected dipole direction at the site is $D = 0, I = 54.5$.

2) Type the following to find out about **cart_hist**:

% **cart_hist -h**

or check Appendix 1.

To compare a set of directions with a known direction, type:

% **cart_hist -d 0 54.5** < **ex4.9** | **plotxy**

(see Figure 4.32).

3) To determine the IGRF direction at the site (see also Example 1.3), type the following:

% **igrf**

1997 0 35 -120

and the computer responds:

14.4 59.5 48925.

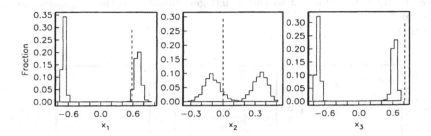

Figure 4.32. Histograms from Example 4.11, using the command: **cart_hist -d 0 54.5** < **ex4.9 | plotxy**. The dashed lines are the cartesian coordinates of the input direction 0, 54.5.

4) The geomagnetic reference field at the sampling site was $D = 14.3, I = 59.5$ so, to compare with the data, type:
% **cart_hist -d 14.4 59.5** <**ex4.9 | plotxy**
(see Figure 4.33).

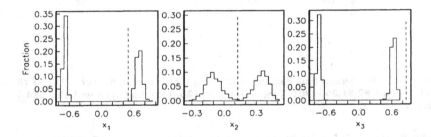

Figure 4.33. Histograms from Example 4.11, using the command: **cart_hist -d 14.4 59.5** < **ex4.9 | plotxy**. The dashed lines are the cartesian coordinates of the input direction 14.4, 59.5.

• Example 4.12
Pretend that a reference pole ($\phi = -140°, \lambda = 75°$) was determined for rocks of the same age as those that yielded the data in Table 4.2. The VGPs that went into the calculation had a reported κ of 30. 1) Assume that the VGPs were Fisher distributed and generate a synthetic data set with the same mean and κ ($N = 100$) using **fishrot**. 2) Convert these data to the expected directions at the sampling site for the last example. 3) Com-

pare the data in Table 4.2 to those just generated. Are they significantly different?
Solution

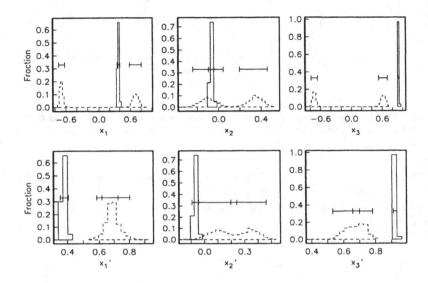

Figure 4.34. Histograms from Example 4.12, using the command: **fishrot -di -140 75 | awk '{print \$2,\$1,35,-120}' | vgp_di > ex4.12b; cart_hist -crb ex4.12b ex4.9 | plotxy**.

Because you do not have access to the original data, you can assume that the VGPs were Fisher distributed and create a synthetic distribution with the same mean direction and κ.

Because the default kappa, N, and seed of **fishrot** are acceptable, we need only specify the VGP position by typing:

% **fishrot -di -140 75 > ex4.12a**

Now we have to enter the site location, rearrange the pole latitude and longitude and pipe it to **vgp_di** (see Example 1.6):

% **awk '{print \$2,\$1,35,-120}' ex4.12a | vgp_di > ex4.12b**

To complete the problem, we must compare the data in **ex4.9** with those in **ex4.12b** using **cart_hist**. We also need to flip the reversed polarity data and plot the 95% confidence bounds.

Type:

% **cart_hist -crb ex4.12b ex4.9 | plotxy**
and look at Figure 4.34. It appears that the data in **ex4.9** are significantly different from the dipole field, the present geomagnetic field and the reference directions.

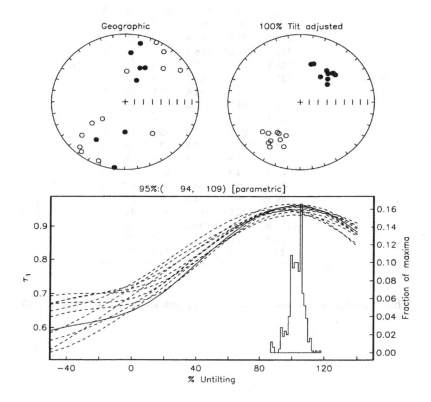

Figure 4.35. Fold test of Example 4.13, using the command: **foldtest -p < ex4.13 | plotxy.**

• **Example 4.13**
Use program **foldtest** to perform a parametric fold test on the data in Table 4.3.
Solution
First put the data into a data file called **ex4.13**. Then, type the following:
% **foldtest -p < ex4.13 | plotxy**

TABLE 4.3. Data for Example 4.13.

Dec.	Inc.	Strike	Dip	κ	N
218.8	-6.3	160.6	50.7	77	5
215.2	32.8	129.6	60.4	23	5
202.3	-6.5	62.4	42.8	49	5
141.3	-40.6	359.4	64.2	41	5
222.7	-10.6	198.1	36.9	46	5
235.2	-44.6	232.3	45.4	37	5
231.0	-57.6	298.1	21.1	77	5
188.1	2.3	54.6	62.1	43	5
2.4	-52.7	290	80	35	5
230.2	-24.7	146.0	8.8	28	5
5.6	27.9	52.4	55.8	57	5
58.1	-12.9	178.7	59.7	70	5
21.3	45.5	226.1	8	17	5
25.3	-3.3	80.0	60.4	71	5
12.9	12.5	50.0	64.6	82	5
46.1	-29.6	134.3	67.7	36	5
35.0	-10.5	116.2	52.1	44	5
180.0	53.6	295.3	73.4	76	5
27.8	42.6	81.1	12	39	5
25.3	61.3	275.8	21.8	69	5

which produces the postscript file **mypost** as shown in Figure 4.35.

Chapter 5

ANALYZING TENSORS

In the previous chapter, we were concerned with magnetic vectors. Magnetic tensors are also tremendously useful in geological studies and have found wide use in studies involving sedimentary, igneous and metamorphic rocks. Such data have applications in determining such varied parameters as paleocurrent directions, degree of paleosol maturity, directions of magma injection, tectonic strain, etc. In this chapter, we will discuss techniques for obtaining and analyzing magnetic tensor data.

5.1. The magnetic susceptibility tensor

The relationship between a small applied magnetic field vector $\mathbf{H}$ and the induced magnetization vector $\mathbf{M}$ can often be approximated by a set of linear equations. Components of the induced magnetization in a given coordinate system whose axes are denoted by $\mathbf{X}_1, \mathbf{X}_2$, and $\mathbf{X}_3$ are related to the applied field by the following linear equations:

$$\begin{aligned}
M_1 &= \chi_{11}H_1 + \chi_{12}H_2 + \chi_{13}H_3 \\
M_2 &= \chi_{21}H_1 + \chi_{22}H_2 + \chi_{23}H_3 \\
M_3 &= \chi_{31}H_1 + \chi_{32}H_2 + \chi_{33}H_3,
\end{aligned} \qquad (5.1)$$

where χ_{ij} are the susceptibility coefficients.

A linear relationship between two vectors can be formulated as a second-order tensor. The constants χ_{ij} are the elements of a second-order, symmetric tensor which is known as the *anisotropy of magnetic susceptibility (AMS) tensor* χ. The set of equations 5.1 can be rewritten in subscript notation as:

$$M_i = \chi_{ij}H_j. \qquad (5.2)$$

Remanences such as TRM, ARM and IRM (Chapter 2) that are acquired in applied fields are only linearly related to the field under special conditions (e.g., low fields), hence the equivalent equations cannot, strictly speaking, be described by a second-order tensor as in equation 5.2. However, many people (e.g., Jackson et al. [1988, 1989] and Lu and McCabe [1993]) have treated the problem in an analogous manner to the tensor problem. The analysis described here would be equally appropriate to remanences as for susceptibility.

Since $\chi_{ij} = \chi_{ji}$, the susceptibility tensor χ defines a symmetric, second-order tensor that has 6 independent matrix elements. For convenience we define a related column matrix s having six elements that are related to the elements of χ by:

$$s_1 = \chi_{11}$$
$$s_2 = \chi_{22}$$
$$s_3 = \chi_{33}$$
$$s_4 = \chi_{12} = \chi_{21}$$
$$s_5 = \chi_{23} = \chi_{32}$$
$$s_6 = \chi_{13} = \chi_{31}.$$

(5.3)

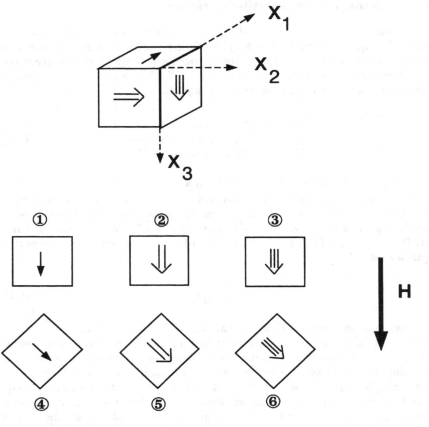

Figure 5.1. Scheme for measuring the six elements of **K**. The sample is placed in six different orientations in a solenoid within a weak applied field **H**.

In many laboratories, magnetic susceptibility is measured by placing a sample in a solenoid with a applied field **H**. The induced magnetization **M** parallel to $\mathbf{H}$ is measured in different orientations. Only s_1, s_2, and s_3 can be directly measured and the terms s_4 to s_6 are only indirectly determined.

We define a set of six values of measured susceptibility K_i which are determined in the six sample positions shown in Figure 5.1. Measurement in position 1 gives $K_1 = s_1$. Similarly, $K_2 = s_2$, and $K_3 = s_3$. But, $K_4 = \frac{1}{2}(s_1 + s_2) + s_4$, $K_5 = \frac{1}{2}(s_2 + s_3) + s_5$, and $K_6 = \frac{1}{2}(s_1 + s_3) + s_6$.

Thus the elements of $\mathbf{s}$ are related to $\mathbf{K}$ in subscript notation by:

$$K_i = A_{ij}s_j, \tag{5.4}$$

where $\mathbf{A}$ is the so-called *design matrix* (see Hext, [1963]).

From the equations for the six measurements shown in Figure 5.1, we can write:

$$\mathbf{A} = \begin{pmatrix} 1 & 0 & 0 & 0 & 0 & 0 \\ 0 & 1 & 0 & 0 & 0 & 0 \\ 0 & 0 & 1 & 0 & 0 & 0 \\ .5 & .5 & 0 & 1 & 0 & 0 \\ 0 & .5 & .5 & 0 & 1 & 0 \\ .5 & 0 & .5 & 0 & 0 & 1 \end{pmatrix}. \tag{5.5}$$

Although there are six measurements and six unknowns, the elements of $\mathbf{s}$ are overdetermined, because the diagonal measurements depend on three parameters. In order to calculate the best-fit values $\bar{\mathbf{s}}$ for $\mathbf{s}$, we can use linear algebra:

$$\bar{\mathbf{s}} = (\mathbf{A}^T\mathbf{A})^{-1}\mathbf{A}^T\mathbf{K} \quad \text{or} \quad \bar{\mathbf{s}} = \mathbf{B}\mathbf{K}, \tag{5.6}$$

where $\mathbf{A}^T$ is the transpose of $\mathbf{A}$. The elements of $\mathbf{B}$ shown in Figure 5.1 are:

$$\mathbf{B} = \begin{pmatrix} 1 & 0 & 0 & 0 & 0 & 0 \\ 0 & 1 & 0 & 0 & 0 & 0 \\ 0 & 0 & 1 & 0 & 0 & 0 \\ -.5 & -.5 & 0 & 1 & 0 & 0 \\ 0 & -.5 & -.5 & 0 & 1 & 0 \\ -.5 & 0 & -.5 & 0 & 0 & 1 \end{pmatrix}. \tag{5.7}$$

In the special case in which $\mathbf{A}$ is a square matrix (as in equation 5.5, $(\mathbf{A}^T\mathbf{A})^{-1}\mathbf{A}^T$ reduces to $\mathbf{A}^{-1}$.

There exists one coordinate system $\mathbf{V}$ (whose axes are the eigenvectors of χ: $\mathbf{V}_1$, $\mathbf{V}_2$, $\mathbf{V}_3$) in which the off-axis terms of χ are zero (Chapter 3). In this coordinate system:

$$\begin{aligned} M_1 &= s_1 H_1 = \chi_{11} H_1 \propto \tau_1 H_1 \\ M_2 &= s_2 H_2 = \chi_{22} H_1 \propto \tau_2 H_2 \\ M_3 &= s_3 H_3 = \chi_{33} H_3 \propto \tau_3 H_3. \end{aligned} \tag{5.8}$$

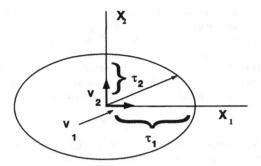

Figure 5.2. The magnitude ellipse of Nye [1957]. The coordinate system is defined by the eigenvectors V_1 and V_2. The lengths of the semi-axes of the ellipse are proportional to the eigenvalues τ_1 and τ_2. The magnitude ellipse is also called the AMS ellipse or ellipsoid in the three-dimensional case.

Following the treatment of the orientation tensor **T** in Chapter 3, we can express the condition of zero off-axis terms as:

$$\chi V = \tau V, \tag{5.9}$$

where τ is the diagonal matrix of the eigenvalues τ_i and **V** contains the components of the eigenvectors:

$$V = \begin{pmatrix} v_{11} & v_{21} & v_{31} \\ v_{12} & v_{22} & v_{32} \\ v_{13} & v_{23} & v_{33} \end{pmatrix}.$$

The solution to equation 5.9 was described in Chapter 3 (see also Press et al. [1986]). The eigenvalues τ_1, τ_2 and τ_3 correspond to the maximum, intermediate, and minimum susceptibility, respectively. These are the susceptibilities along the principal, major and minor eigenvectors V_1, V_2, and V_3, respectively. Scaling χ by its trace yields values for τ that sum to unity. V_1, V_2 and V_3 are sometimes referred to as K_{max}, K_{int}, and K_{min}, respectively in the literature.

[See Example 5.1]

When the coordinate system of the susceptibility data is defined by the eigenvectors, then the components of magnetization M_i satisfy the following:

$$\frac{M_1^2}{\tau_1^2} + \frac{M_2^2}{\tau_2^2} + \frac{M_3^2}{\tau_3^2} = 1. \tag{5.10}$$

When the coordinate system is referred to the eigenvectors, as illustrated (for the 2D case) in Figure 5.2, the surface described by equation 5.10 traces an ellipsoid termed the *magnitude ellipsoid* by Nye [1957] whose semi-axes directed along **V** have lengths that are proportional to the τ_i. We will refer to this ellipsoid in the following as the anisotropy of magnetic susceptibilty (AMS) ellipsoid.

[See Example 5.2]

Many publications list AMS data in terms of the eigenvalues and eigenvectors, collectively referred to as eigenparameters. We need a way to transform eigenparameters back into matrix elements. This can be done using tricks from linear algebra. It follows from equation 5.9 that

$$\chi = \mathbf{V}\tau\mathbf{V}^T, \qquad (5.11)$$

where $\mathbf{V}^T$ is the transpose of **V**. Furthermore, since anisotropy data are obtained from samples with an arbitrary orientation, these data must be transformed into geographic and tilt adjusted coordinates. The procedure is analogous to the transformation for unit vectors described in Chapter 3 and is achieved by the following matrix multiplication:

$$v'_{ij} = a_{ik}a_{jl}v_{kl}, \qquad (5.12)$$

where v'_{ij} are the components of the eigenvectors in the transformed coordinate system and the as are the appropriate direction cosines.

[See Example 5.3]

The eigenparameters of the susceptibility tensor are related to the statistical alignment of dia-, para-, and/or ferromagnetic phases within the rock and the AMS ellipsoid can be used to describe the magnetic fabric of the rock. Much of the interpretation of AMS data in the literature revolves around an assessment of directions of principal axes and relative magnitudes of the eigenvalues.

There is a bewildering variety of conventions for describing the relationships among the three eigenvalues (see, e.g., Table 1.1 of Tarling and Hrouda [1993] and Table 5.1). A practical initial classification scheme can be made with the following rules: when ($\tau_1 \simeq \tau_2 \simeq \tau_3$), the shape is a sphere; when ($\tau_1 \simeq \tau_2 > \tau_3$), it is oblate. The shape is prolate when ($\tau_1 > \tau_2 \simeq \tau_3$), and, finally, the anisotropy ellipsoid is triaxial when ($\tau_1 > \tau_2 > \tau_3$). Because there are nearly always three distinct values of τ, it is a statistical problem to decide whether the eigenvalues from a given data set are significantly different from one another.

Making only six measurements allows calculation of the eigenparameters, but gives no constraints for their uncertainties. We would like to ask questions such as the following:

1) Is a particular axis parallel to some direction? Is V_3 vertical as for a primary sedimentary fabric? Is V_1 parallel to some lineation such as elongated vesicles in volcanic dikes, or deformed ooids in strained rocks?
2) Are two sets of eigenvectors distinct? Are data from two sides of a dike margin imbricated, allowing interpretation of flow direction? Has progressive strain rotated the rock fabrics?
3) What is the shape of the AMS ellipsoid? Are the eigenvalues distinct? Is the fabric oblate, as for primary sedimentary rocks? Does the shape change as a result of progressive deformation in metamorphic rocks?

In order to address questions such as these, we need some sort of confidence intervals for the eigenparameters; hence we need to make multiple measurements and we need a means of translating the measurements into uncertainties in AMS data. The principles of error analysis for anisotropy measurements were originally laid out by Hext [1963]. Jelinek [1976, 1978] developed the ideas further and what we describe as *linear perturbation analysis* (LPA) has become the standard method of estimating uncertainties (see e.g., Tarling and Hrouda [1993]).

5.2. Linear Perturbation Analysis (LPA)

According to Hext [1963], each measurement K_i has an unknown measurement "error":

$$K_i = A_{ij}s_j + \delta_i. \tag{5.13}$$

The residual sum of squares S_o is:

$$S_o = \sum_i \delta_i^2, \tag{5.14}$$

and the estimated variance is:

$$\sigma^2 = S_o/n_f. \tag{5.15}$$

n_f is the number of degrees of freedom, given by $N - 6$ where N is the number of measurements. The covariance matrix of s is given by:

$$\sigma^2(\mathbf{A}^T\mathbf{A})^{-1}. \tag{5.16}$$

Following Box and Hunter [1957], Hext [1963] developed measurement schemes that have nearly spherical variance functions on a unit sphere. Measurement schemes that are evenly spaced over the unit sphere are termed *rotatable designs*. Hext [1963] proposed several rotatable measurement schemes having 12 and 24 measuring positions. Jelinek [1976] proposed a rotatable measurement scheme with 15 measuring positions, as illustrated in Figure 5.3. This is the procedure recommended in the manual distributed with the popular Kappabridge susceptiblity instruments. In the 15 measurement case shown in Figure 5.3, the design matrix is:

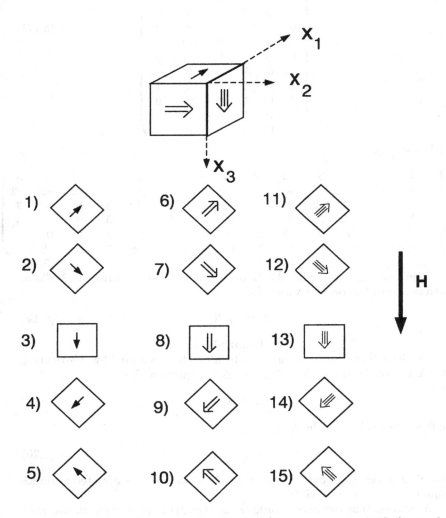

Figure 5.3. The 15 position scheme of Jelinek [1976] for measuring the AMS of a sample.

$$\mathbf{A} = \begin{pmatrix} .5 & .5 & 0 & -1 & 0 & 0 \\ .5 & .5 & 0 & 1 & 0 & 0 \\ 1 & 0 & 0 & 0 & 0 & 0 \\ .5 & .5 & 0 & -1 & 0 & 0 \\ .5 & .5 & 0 & 1 & 0 & 0 \\ 0 & .5 & .5 & 0 & -1 & 0 \\ 0 & .5 & .5 & 0 & 1 & 0 \\ 0 & 1 & 0 & 0 & 0 & 0 \\ 0 & .5 & .5 & 0 & -1 & 0 \\ 0 & .5 & .5 & 0 & 1 & 0 \\ .5 & 0 & .5 & 0 & 0 & -1 \\ .5 & 0 & .5 & 0 & 0 & 1 \\ 0 & 0 & 1 & 0 & 0 & 0 \\ .5 & 0 & .5 & 0 & 0 & -1 \\ .5 & 0 & .5 & 0 & 0 & 1 \end{pmatrix}, \tag{5.17}$$

and $\mathbf{B} = \frac{1}{20} \times$

$$\begin{pmatrix} 3 & 3 & 8 & 3 & 3 & -2 & -2 & -2 & -2 & -2 & 3 & 3 & -2 & 3 & 3 \\ 3 & 3 & -2 & 3 & 3 & 3 & 3 & 8 & 3 & 3 & -2 & -2 & -2 & -2 & -2 \\ -2 & -2 & -2 & -2 & -2 & 3 & 3 & -2 & 3 & 3 & 3 & 3 & 8 & 3 & 3 \\ -5 & 5 & 0 & -5 & 5 & 0 & 0 & 0 & 0 & 0 & 0 & 0 & 0 & 0 & 0 \\ 0 & 0 & 0 & 0 & 0 & -5 & 5 & 0 & -5 & 5 & 0 & 0 & 0 & 0 & 0 \\ 0 & 0 & 0 & 0 & 0 & 0 & 0 & 0 & 0 & 0 & -5 & 5 & 0 & -5 & 5 \end{pmatrix}.$$

$$\tag{5.18}$$

Using equation 5.6 with the **B** matrix defined in equation 5.18 allows calculation of the best-fit values for **s**:

$$\bar{s}_i = B_{ij}K_j. \tag{5.19}$$

[See Example 5.4]

The best-fit values for **K** are $\bar{\mathbf{K}}$, which can be calculated by substituting the **A** matrix from equation 5.17 for **A** in equation 5.4:

$$\bar{K}_i = A_{ij}\bar{s}_j.$$

Now we can calculate the δ_i by:

$$\delta_i = K_i - \bar{K}_i, \tag{5.20}$$

and S_o is given by equation 5.14. Here $N = 15$, so the estimated variance (equation 5.15) is $\sigma^2 = S_o/9$.

Assuming that the uncertainties in **K** (the δ_i) have zero mean, and they are uncorrelated, normally distributed, and small (so that the products of

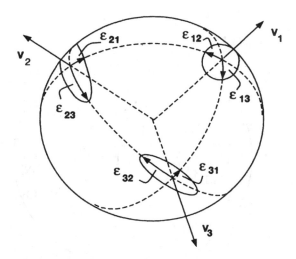

Figure 5.4. Relationship of the uncertainty ellipses (calculated by LPA for AMS data) to the principal axes. The major and minor semi-axes of the uncertainty ellipses are oriented along the axes defined by thé eigenvectors.

uncertainties can be neglected), Hext [1963] proposed that approximate 95% confidence ellipses for the eigenvectors can be found by LPA.

First, we assume that the uncertainties in the eigenvectors are in a plane that is tangent to the unit sphere. We further assume that they belong to a two-dimensional normal distribution with semi-axes that are aligned along the V_i. The ellipse with semi-axes ϵ_{ij} that outline a 95% confidence region in this plane is then projected onto the sphere (Figure 5.4).

There are two approaches for estimating confidence regions by LPA: a "short-cut" outlined by Hext [1963], and a more intensive calculation that was sketched by Hext [1963], but which was described more fully by Jelinek [1978]. We will call these two methods the "Hext" method and the "Jelinek" method, respectively. The Hext method is implemented in the popular measurement program ANISO10 as described by Jelinek [1976] and which is distributed with the Kappabridge instrument. It is unclear which method most investigators actually use when they refer to the "Jelinek method."

5.3. The Hext method of LPA

The procedure for calculating confidence regions using the Hext [1963] method is as follows:

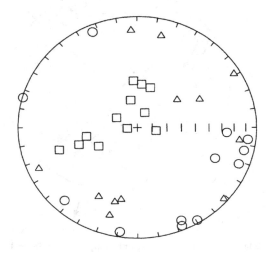

Figure 5.5. Lower hemisphere projection of directions of V_1 (squares), V_2 (triangles), and V_3 (circles) from the margin of a volcanic dike in the Troodos Ophiolite.

- Calculate the matrices $\bar{s}$, $\bar{K}$, and the δ_i from the measured values of K (see Example 5.4).
- Calculate the eigenvectors V and eigenvalues τ of $\bar{s}$ (see Example 5.1).
- Calculate σ by equations 5.14 and 5.15. (see Example 5.4).
- The confidence regions are outlined by ellipses along semi-axes ϵ_{ij} aligned with the eigenvectors. The i subscripts refer to the axis on which the ellipse is attached and the j subscripts refer to the axis to which it points. Thus, ϵ_{12} is the semi-axis that defines the confidence region of V_1 directed toward V_2 (Figure 5.4).

The three unique semi-angles of the confidence ellipses ϵ_{ij} are calculated by:

$$\epsilon_{12} = \tan^{-1}[f\sigma/2(\tau_1 - \tau_2)]$$
$$\epsilon_{23} = \tan^{-1}[f\sigma/2(\tau_2 - \tau_3)]$$
$$\epsilon_{13} = \tan^{-1}[f\sigma/2(\tau_1 - \tau_3)]$$
$$\epsilon_{21} = \epsilon_{12}$$
$$\epsilon_{32} = \epsilon_{23}$$
$$\epsilon_{31} = \epsilon_{13},$$

(5.21)

where

$$f = \sqrt{2(F_{(2,n_f);(1-p)})},$$

and where $F_{(2,n_f)}$ is the value from the F table, with 2 and n_f degrees of freedom, at the p probability level. The value of $F_{(2,n_f)}$ for $N = 15$ measurements ($n_f = 9$) at the 95% level of confidence ($p = .05$) is 4.26 and so $f = 2.92$.

We turn now to the problem of calculating confidence intervals for the eigenvalues. In practice, there will almost always be three distinct values of τ returned from an eigenvalue calculation. But, when are these values statistically distinct? First, one might ask if the ellipsoid is significantly different from a sphere. LPA allows calculation of F statistics and comparison with values in F tables; in this way one can test if the data are isotropic (F) ($\tau_1 = \tau_2 = \tau_3$), if $\tau_1 = \tau_2$ (F_{12}), or if $\tau_2 = \tau_3$ (F_{23}). We calculate the F statistics by:

$$F = 0.4(\tau_1^2 + \tau_2^2 + \tau_3^2 - 3\chi_b^2)/\sigma^2$$
$$F_{12} = 0.5((\tau_1 - \tau_2)/\sigma)^2$$
$$F_{23} = 0.5((\tau_2 - \tau_3)/\sigma)^2,$$

(5.22)

where the bulk susceptibility χ_b is given by:

$$\chi_b = (\bar{s}_1 + \bar{s}_2 + \bar{s}_3)/3.$$

(5.23)

[See Example 5.5]

The critical value for F is 3.4817 for 95% confidence (for F_{12} and F_{23}, it is 4.2565). F values below these critical values do not allow rejection of the hypothesis of isotropy or rotational symmetry, respectively.

5.4. Multiple samples

In the foregoing discussion, we outlined a way of analyzing data from a sample. Often, we are interested in determining the average behavior from multiple samples from a given study area. To do this, we must average multiple samples from each site and somehow combine the data into average eigenparameters and their uncertainty. There is no *a priori* reason why the

Hext analysis cannot be extended to multiple samples from a given site, as long as the principal assumptions concerning noise are valid. The procedure is as follows:

• Find the s matrix from the N_m measurements for each of N_s samples. In the following, s_{il} is the ith matrix element ($i = 1 \to 6$) of the lth sample ($l = 1 \to N_s$). Rotate these into a common coordinate system (see equation 5.1).

• Normalize the data from each sample by ($s_1 + s_2 + s_3$) so that all samples are given equal weight.

• Use these data to calculate a mean $\bar{S}$ of all the individual estimates of s in which a given matrix element of $\bar{S}_i$ is calculated by summing the N_s estimates of s_i, i.e.,

$$\bar{S}_i = \sum_l s_{il}. \qquad (5.24)$$

• Now calculate the expectation values of **K** ($\bar{K}$), and the δ_i for each element of $\bar{S}$. Calculate Σ (the standard deviation for the site) as σ was calculated for an individual sample before, using these values for δ_i and substituting $6N_s$ for N.

• Calculate the eigenparameters and confidences, as before. Here, however, $n_f = 6N_s - 6$.

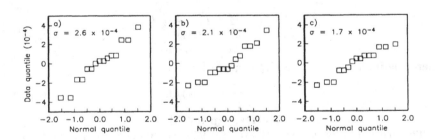

Figure 5.6.   Q-Q plots of the 15 δ values for three representative samples (a-c) from Figure 5.5. The samples were measured using the 15 measurement protocol of Jelinek [1976], hence $N = 15$. The Q-Q plots are approximately linear which suggests that the δs are approximately normally distributed.

[See Examples 5.6 and 5.7]

As pointed out by Jelinek [1978], the approximations used in the Hext [1963] method limit its application to data with very small variances. Measurements of single samples on highly sensitive instruments usually fall in this category, but data from less sensitive instruments or data from different samples from the same study area may not. It was for this reason that Jelinek [1978] elaborated on the approach of Hext [1963] and described the following so-called *Jelinek* method of LPA.

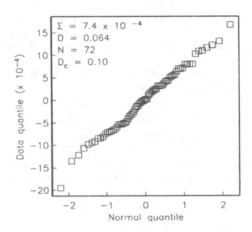

Figure 5.7.   Q-Q of δ values for the sample data from Figure 5.5. The Kolmogorov-Smirnov parameter D is lower than the critical value D_c for N, hence a normal distribution for the δ values cannot be excluded.

5.5.   The Jelinek method of LPA

Taking the l estimates for (normalized) **S** and the average $\bar{\mathbf{S}}$ as before, we now calculate the covariance matrix **C** for a 6 x 6 matrix whose elements are given by:

$$C_{jk} = N_s^{-1} \sum_l (S_{lj} - \bar{S}_j)(S_{lk} - \bar{S}_k). \tag{5.25}$$

The **C** matrix can be transformed into **C'**, the coordinates of the eigenvectors, by matrix manipulation:

$$\mathbf{C'} = \mathbf{G}\mathbf{C}\mathbf{G}^T, \tag{5.26}$$

where **G** is the matrix:

$$\mathbf{G} =$$

$$\begin{pmatrix} v_{11}^2 & v_{21}^2 & v_{31}^2 & 2v_{11}v_{21} & 2v_{21}v_{31} & 2v_{31}v_{11} \\ v_{12}^2 & v_{22}^2 & v_{32}^2 & 2v_{12}v_{22} & 2v_{22}v_{32} & 2v_{32}v_{12} \\ v_{13}^2 & v_{23}^2 & v_{33}^2 & 2v_{13}v_{23} & 2v_{23}v_{33} & 2v_{33}v_{13} \\ v_{11}v_{12} & v_{21}v_{22} & v_{31}v_{32} & v_{11}v_{22}+v_{21}v_{12} & v_{21}v_{32}+v_{31}v_{22} & v_{31}v_{12}+v_{11}v_{32} \\ v_{12}v_{13} & v_{22}v_{23} & v_{32}v_{33} & v_{12}v_{23}+v_{22}v_{13} & v_{22}v_{33}+v_{32}v_{23} & v_{32}v_{13}+v_{12}v_{33} \\ v_{13}v_{11} & v_{23}v_{21} & v_{33}v_{31} & v_{13}v_{21}+v_{23}v_{11} & v_{23}v_{31}+v_{33}v_{21} & v_{33}v_{11}+v_{13}v_{31} \end{pmatrix}.$$

$$(5.27)$$

There is a 2 x 2 covariance matrix $\mathbf{W}_i$ for each eigenvector $\mathbf{V}_i$ which describes the variability of that axis:

$$\mathbf{W}_i = \begin{pmatrix} \dfrac{C'_{i+3,i+3}}{(\tau_i-\tau_j)^2} & \dfrac{C'_{i+3,k+3}}{(\tau_i-\tau_j)(\tau_i-\tau_k)} \\ \dfrac{C'_{i+3,k+3}}{(\tau_i-\tau_j)(\tau_i-\tau_k)} & \dfrac{C'_{k+3,k+3}}{(\tau_i-\tau_k)^2} \end{pmatrix}, \qquad (5.28)$$

where the indices i, j and k are cyclic permutations of the numbers 1, 2, 3.

Each $\mathbf{W}_i$ has two eigenvalues λ_m. These relate to the semi-axes ϵ_{im} by:

$$\epsilon_{im} = \tan^{-1}(\sqrt{f\lambda_m}), \qquad (5.29)$$

where $f = 2(N-1)[N(N-2)]^{-1}F_{2,N-2}$.

[See Example 5.8]

While there is a test for overall anisotropy (similar to the F test described in the Hext method), there are no analogous tests for distinguishing between different eigenvalues. This is a handicap of the Jelinek method which we address later in the chapter. First, it is worthwhile considering when and whether the Jelinek or Hext methods are appropriate.

5.6. When are data suitable for LPA?

The assumptions for using the techniques outlined in the foregoing are that the uncertainties in the measurements have zero mean, are normally distributed, and are small. While measurement error using modern equipment is likely to be quite small, data from a collection of samples often do not conform to these restrictive assumptions. In particular, the δ values are often large.

The eigenvector data in Figure 5.5 were obtained by analyzing numerous individually oriented samples from one of the quenched margins of a dike in the Troodos Ophiolite on Cyprus (see Tauxe et al. [1998]). We plot the eigenvectors on an equal area net using the lower hemisphere projection and follow the convention that the $\mathbf{V}_1$s are squares, $\mathbf{V}_2$s are triangles, and $\mathbf{V}_3$s are circles (Ellwood et al. [1988]).

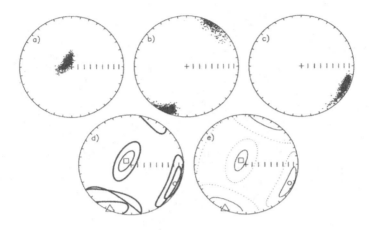

Figure 5.8. Comparison of bootstrapped and LPA confidence ellipses for principal eigenvectors of the AMS data shown in Figure 5.5. a) Equal area projection of principal eigenvectors (V_1) of 500 para-data sets. b) Same as a) for the major eigenvectors (V_2). c) Same as a) for the minor eigenvectors (V_3). d) The inner ellipses are the approximate 95% confidence ellipses estimated assuming a Kent distribution for the bootstrapped eigenvectors shown in a-c. The outer ellipses are the same for a site parametric bootstrap (see text). e) Confidence ellipses calculated using the Jelinek method (solid line) and the Hext method (dotted).

The data are rather typical of data taken from a single homogeneous body of rock. Representative values of δ for these samples calculated from 15 measurements (see Figure 5.3) are plotted as a Q-Q plot in Figure 5.6 against a normal distribution (Chapter 4). The data shown in Figure 5.6 have zero mean and, based on the linearity of the Q-Q plots are, likely to be normally distributed. Moreover, the magnitude of the δ values is rather small with σ's of a few parts in 10,000.

In Figure 5.7 we plot the δ data for calculated for the entire site. Here the Σ is several times the σs for invidual sample measurements. Based on the linearity of the Q-Q plots and the Kolmogorov-Smirnov test, the δ values are also consistent with a normal distribution. Many data sets, particularly those in which a site does not constitute a single, homogeneous rock body, produce δ distributions that are neither normally distributed, nor small. Moreover, Constable and Tauxe [1990] showed that, in general, δs from AMS data calculated for multiple samples (that must be normalized by their trace) will not generally be normally distributed. Hence, data incorporating

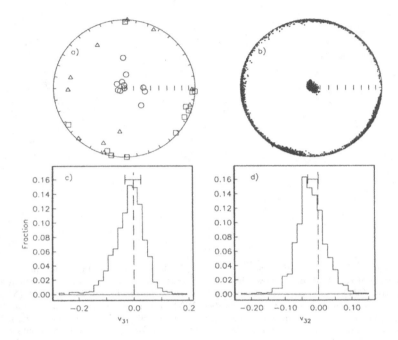

Figure 5.9.   a) AMS data from Cretaceous carbonate limestones in Italy (the Scaglia Bianca Formation) in tilt adjusted coordinates. a) Lower hemisphere projections of the principal V_1 (squares), major V_2 (triangles), and minor V_3 (circles) eigenvectors. b) Bootstrapped eigenvectors from para-data sets of the data in a). c) Histogram of the v_{31} components of V_3 from b) with bounds containing 95% of the components. The zero value expected from a vertical direction is shown as a dashed line. d) same as c) but for the v_{32} component. Since both v_{31} and v_{32} are indistinguishable from zero, V_3 is vertical at the 95% level of confidence.

multiple samples are often not amenable to linear perturbation analysis.

5.7. Bootstrap confidence ellipses

In analogy to the bootstrap for unit vectors, Constable and Tauxe [1990] developed a bootstrap for anisotropy data. As in Chapter 4, we first take a number of randomly selected para-data sets of the data from those shown in Figure 5.5. The eigenparameters of the bootstrapped average $\bar{S}$ matrices are then calculated. Such bootstrapped eigenvectors are shown in the equal

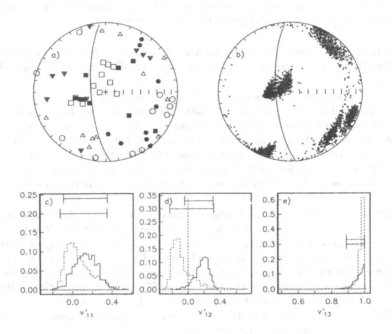

Figure 5.10. AMS data obtained from samples from the margins of a nearly vertical dike in the Troodos Ophiolite. a) Equal area projection of principal (squares), major (triangles), and minor (circles) eigenvectors. Data from the Eastern (Western) margins are shown as open (solid) symbols. The dike trace is shown as a great circle. b) Bootstrapped eigenvectors from the data shown in a). c) Histogram of cartesian coordinates v'_{11} of the bootstrapped principal eigenvectors. These have been rotated into "dike coordinates" (see text). d) Same as c) but for v'_{12}. The dike plane (dashed line) is centered with respect to the $\mathbf{X}_2$ direction. The v_{12} componenent from neither margin is distinct from the dike plane or from each other, based on the 95% confidence bounds drawn above the histograms. e) Same as c) but for v'_{13}.

area projections in Figure 5.8(a-c).

A non-parametric confidence region for the bootstrapped distributions shown in Figure 5.8 could be drawn as a contour line enclosing 95% of the bootstrapped eigenvectors. Because it is often useful to characterize the average uncertainties with a few parameters (for example, to put them in a data table), we can proceed as with the unit vectors and assume some sort of distribution for the eigenvectors. As in Chapter 4, we use the Kent distribution. Our method for estimated 95% confidence ellipses guided by bootstrap distributions of eigenvectors is as follows:

- Randomly select a para-data set from the list of N_s s data and calculate $\bar{S}_k$ for each para-data set, where $\bar{S}_k$ is the $\bar{S}$ from the kth para-data set.
- Calculate the eigenparameters V and τ for each $\bar{S}$.
- Repeat the first two steps N_b (say 1000) times, such that v_{1k} and τ_{1k} represent the principal eigenparameters of the kth para-data set.
- Use the method outlined in Chapter 4 for estimating the confidence ellipses for each eigenvector. Although the confidence ellipses are calculated separately for each eigenvector, the eigenvectors are constrained to be those of $\bar{S}$ (i.e. the same as for the Hext and Jelinek methods).

By analogy with the bootstrap for unit vectors and the fold test, we can also perform two types of parametric bootstraps. The first flavor, or the *sample parametric bootstrap* proceeds as follows: After randomly selecting a particular sample for inclusion, each element of s is replaced by a simulated element drawn from a normal distribution having the σ as calculated for the sample data (see Example 4.1). This Monte Carlo type simulation assumes that the measurement uncertainties are normal, which as discussed earlier, is likely to be the case. If instrument noise is significant, then the sample parametric bootstrap is an important tool.

Because the δ_i data from homogeneous rock bodies are often normally distributed, we can also perform a parametric bootstrap at the level of the site (the *site parametric bootstrap*). This is done by drawing para-data sets as before, but replacing individual elements of s with simulated data drawn from normal distributions with Σ calculated from the data for an entire site. This procedure goes a long way toward calculating realistic confidence intervals from sites with too few samples (see Tauxe et al. [1998]).

[See Example 5.9]

In Figure 5.8 we compare the confidence ellipses obtained from the simple and parametric bootstraps with those obtained by the Hext and Jelinek approaches. The best-fit eigenvectors in the bootstrap are orthogonal; indeed the average eigenparameters are identical in LPA and bootstrapping. The differences between the bootstrap and LPA methods come in the size, shape and orientation of the uncertainty ellipses. First, in the linear perturbation calculation there are but three independent semi-axes of the uncertainty regions and these are oriented along the eigenvectors. This is not true for the bootstrap method which gives six independent semi-axes that are not constrained to by the eigenvectors, but are allowed to "follow the data".

5.8. Comparing mean eigenvectors with other axes

Now we will consider whether a particular axis is distinct from a given direction or another eigenvector. For example, we may wish to know if a given data set from a series of sediments has a vertical minor eigenvector as would be expected for a primary sedimentary fabric (Chapter 6). In Figure 5.9a we show AMS data from samples taken from the Scaglia Bianca Formation (Cretaceous white limestones) in the Umbrian Alps of Italy. They have been rotated into tilt adjusted coordinates; hence the bedding pole is vertical. Instead of plotting the 95% confidence ellipses, which all require unnecessary parametric assumptions, we show the bootstrap eigenvectors in Figure 5.9b. The smear of points certainly covers the vertical direction consistent with a vertical direction for V_3. To make the test at a given level of confidence (say 95%), we can employ the method developed for unit vectors in which the eigenvector of choice (here V_3) is converted to cartesian coordinates and sorted. Then the bootstrapped 95% confidence bounds can be directly compared with the expectation value. For a direction to be vertical, both the x_1 and x_2 components must be zero. We plot v_{31} and v_{32} as histograms, with the 95% confidence bounds shown above in Figure 5.9. The expected value of zero is shown by dashed lines. Because zero is included within the confidence intervals, these data have a direction of V_3 that cannot be distinguished from vertical at the 95% level of confidence.

Another question that often arises is whether eigenvectors from two sets of AMS data can be distinguished. For example, are the V_1 directions from data sets collected from two margins of a dike different from one another and on opposite sides of the dike plane as expected from anisotropy controlled by crystal imbrication (Chapter 6). In Figure 5.10a, we show the eigenvectors of AMS data from samples obtained from both quenched margins of a nearly vertical north-south trending dike in the Troodos Ophiolite in Cyprus. The bootstrapped eigenvectors are shown in Figure 5.10b. In order to address the problem of whether the V_1s are distinct from the dike plane, we first rotate them into *dike coordinates* (whereby the dike pole is parallel to X_2 and direction of dip is parallel to X_3). Then the question of whether the V_1 direction is distinct reduces to whether the v_{12} components can be distinguished from zero (the dike plane). In Figure 5.10c, we show the histogram of the cartesian components of V_1 and the 95% confidence intervals of the two data sets. As we can see from the overlapping confidence bounds, the data are neither distinct from the dike margin, nor from each other.

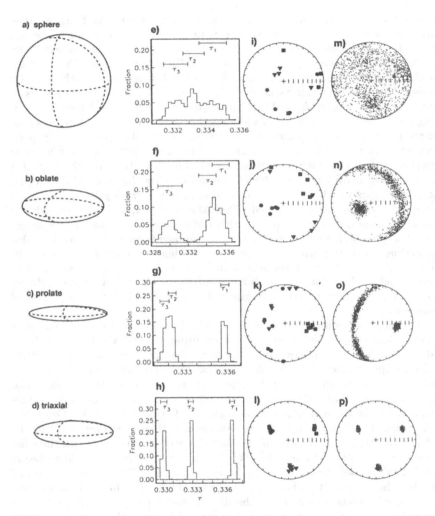

Figure 5.11. Determination shape of AMS ellipsoids using the bootstrap. a)-d) Magnitude ellipsoids. e)-h) Histograms of the bootstrapped eigenvalues associated with the eigenvectors plotted in i)-l) (same conventions as in Figure 5.5). The bounds containing 95% of each eigenvalue are shown above the histograms. m)-p) Bootstrapped eigenvectors.

5.9. Shape

There are innumerable ways of plotting and characterizing shapes of AMS ellipsoids in the literature. We will focus here on a few and discuss how bootstrapping could be helpful in providing a means for discriminating differences in data sets and so on. We list some popular so-called "shape parameters" in Table 5.1.

Many researchers use the "total anisotropy" parameter of Owens [1974]. This has the uncomfortable property of ranging up to 300%; hence, we prefer the parameter called here the % anisotropy of Tauxe et al. [1990] as this ranges from 0 - 100%. The so-called "corrected anisotropy" of Jelinek [1981] has several definitions in the literature (compare for example Borradaile [1988] with Jelinek [1981]); we have used the original definition of Jelinek [1981].

With the plethora of parameters comes a host of plotting conventions. We will consider five types of plots here: histogram of bootstrapped eigenvalues, the Flinn diagram (F versus L) after Flinn [1962], the Ramsay diagram (F' versus L') after Ramsay [1967], the Jelinek diagram (P' versus T) after Jelinek [1981], and the ternary projection (see Woodcock [1977] and Tauxe et al. [1990]). The Flinn, Ramsay, and Jelinek diagrams are shown in Figure 5.12 and the ternary projection is shown in Figure 5.13.

The Flinn and Ramsay diagrams are very similar, but the Ramsay plot has the advantage of having a zero minimum as opposed to starting at 1.0 as in the Flinn diagram. Both are essentially polar plots, with radial trajectories indicating increasing anisotropy. Shape is reflected in the angle, with "oblate" shapes above the line and "prolate" shapes below.

It is important to remember that, in fact, only points along the plot axes themselves are truly oblate or prolate and that all the area of the plot is in the "triaxial" region. Because of statistical uncertainties, samples that plot in this region may fail the F_{12} or F_{23} tests of Hext and be classifiable as "oblate" or "prolate". In general, however, only a narrow zone near the axes can be considered oblate or prolate, so these terms are often used loosely.

The Jelinek diagram is more cartesian in nature than the Flinn or Ramsay plots. "Corrected" anisotropy increases along the horizontal axis and shape reflected in the vertical axis. There is no real advantage to using the highly derived P' and T parameters over the Ramsay or Flinn plots. Nonetheless they are quite popular (Tarling and Hrouda [1993]).

In the ternary projection, there are actually three axes (see Figure 5.13a). The projection can be plotted as a normal X-Y plot by using the E' and R parameters listed in Table 5.1 (see Figure 5.13b).

In none of the various types of plots just discussed are the horizontal

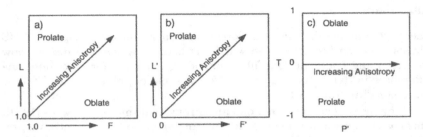

Figure 5.12. Properties of various AMS diagrams: a) Flinn, b) Ramsay and c) Jelinek.

and vertical axes independent of one another, but all the diagrams reflect the essence of the ellipsoid shape. Unlike the histogram with bootstrap confidence intervals, it is not possible to determine whether the various eigenvalues or ratios thereof can be distinguished from one another in a statistical sense.

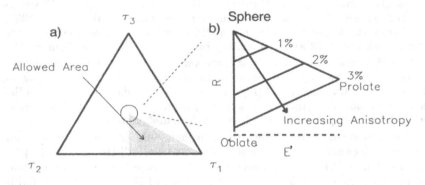

Figure 5.13. Properties of the Ternary diagram: a) there are three axes with limits of τ_1, τ_2, τ_3. Because of the constraint that $\tau_1 > \tau_2 > \tau_3$, only the shaded region is allowed. This is bounded at the top by a sphere when all three eigenvalues are equal, to the bottom left by a disk and to the bottom right by a needle. Geological materials generally have a low degree of anisotropy and plot close to the sphere. This region is enlarged in b) which illustrates how the ternary projection can be plotted as E' versus R and how shape (oblate, prolate, sphere) and percent anisotropy appear on the diagram.

All discussions of the "shape" of the AMS ellipsoid revolve around the relationships between the various eigenvalues. The first question to consider is whether these can be distinguished in a statistical sense. The Hext version of linear perturbation analysis has the ability to check for significance of the anisotropy (using the F parameters). However, the approximations involved in the Hext method make it inappropriate for most data sets involving more than one sample. Bootstrapping allows for testing the significance of the differences in eigenvalues and the less restrictive assumptions allow bootstrap tests to be applied more widely.

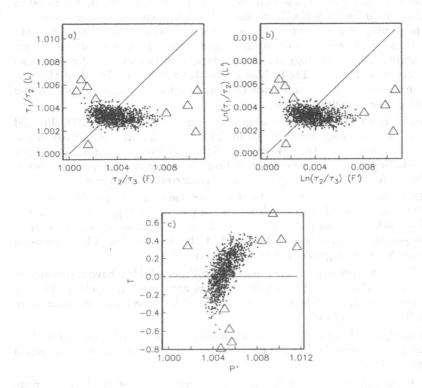

Figure 5.14. Shape parameters from the data whose eigenvectors are plotted in Figure 5.5 shown in three projections: a) Flinn, b) Ramsay, and c) Jelinek. Data from individual samples are shown as triangles, dots are average values for bootstrapped para-data sets.

The simplest means of determining whether two sets of eigenvalues are distinct from one another requires the assumption that the bootstrapped eigenvalues are normally distributed. Here we calculate the standard deviations of the populations of eigenvalues. While the eigenvalues may often satisfy the requirement of normal distribution, they equally often may not.

Hence, we desire a less restrictive way of deciding whether eigenvalues are distinct.

[See Examples 5.10-5.13]

One alternative way of checking whether eigenvalues can be discriminated is quite similar to the bootstrap test for common mean described in Chapter 4. In Figure 5.11a-d, we sketch the four shape categories defined in the beginning of the chapter. The eigenvectors calculated for samples from four sites are shown in Figure 5.11i-l. Bootstrapped eigenvectors are shown in Figure 5.11m-p. Histograms of the associated bootstrapped eigenvalues are shown in Figure 5.11e-h. There are three "humps" in all of the histograms, but the 95% confidence bounds provide a means for quantifying the shape tests defined earlier.

In Figure 5.11a, we illustrate the behavior of data from an AMS ellipsoid that is essentially spherical in shape. The three eigenvalues plotted in the histogram (Figure 5.11e) have overlapping confidence intervals, hence they are indistinguishable. The corresponding bootstrapped eigenvectors shown in Figure 5.11m plot in a cloud with no preferred orientations.

In Figure 5.11b we show data characteristic of an oblate ellipsoid. The smallest eigenvalue is distinct from the other two in Figure 5.11f, but the 95% confidence bounds of τ_2 overlap those of τ_1. The V_3 eigenvector is consequently reasonably well defined, but the distribution of bootstrapped V_2 and V_1 form a girdle distribution (Figure 5.11n).

The data from a prolate ellipsoid (see Figure 5.11c) have a distinct τ_1 distribution (Figure 5.11g), while τ_2 and τ_3 are clumped together. The V_1 directions are nicely defined, but the V_2 and V_3 directions are smeared in a girdle (Figure 5.11o)).

Finally, the triaxial case is shown in Figure 5.11d. All three eigenvalues are distinct (Figure 5.11h) and the corresponding eigenvectors well grouped (Figure 5.11p).

The histogram method illustrated in Figure 5.11 is most appropriate for classifying shape characteristics of a relatively homogeneous set of samples. However, it may not be ideal for examining trends in behavior among samples or data sets. For example, one may wish to show the progressive change in shape and degree of anisotropy as a function of metamorphism (e.g., Chapter 6). In such a case, one of the other plots (Flinn, Ramsay, Jelinek, or ternary) may serve better.

[See Example 5.13]

By way of illustrating various plots, we show the data from Figure 5.5 on the Flinn, Ramsay and Jelinek type plots. We plot both the individual sample data (triangles) and the bootstrapped averages for the whole data set (dots). The dots give a sense of the statistical variability of the average data set and can be used for the purposes of discriminating among data sets in a statistical way (Chapter 6).

While there is no "right" way to plot eigenvalue data, it seems that there is a clear advantage of the histogram approach. The degree of anistropy is reflected directly in the "spread" between τ_1 and τ_3 (see % anisotropy in Table 5.1). The shape is reflected in the relationships among the three eigenvalues which can be quantitatively assessed. Finally, if data from different studies are placed on the same horizontal scale, trends in the data sets are easily observed.

TABLE 5.1. Assorted anisotropy parameters.

Parameter (Reference)	Equation
Bulk Susceptibility (see text)	$\chi_b = (s_1 + s_2 + s_3)/3$
Normalized eigenvalues (see text)	$\tau_1 + \tau_2 + \tau_3 = 1$
Log eigenvalues (Jelinek [1981])	$\eta_1 = \ln \tau_1; \eta_2 = \ln \tau_2; \eta_3 = \ln \tau_3$
Log mean susceptibility (Jelinek [1981])	$\bar{\eta} = (\eta_1 + \eta_2 + \eta_3)/3$
Magnitude of Anisotropy:	
% Anisotropy (Tauxe et al. [1990])	$\%h = 100(\tau_1 - \tau_3)$
"Total" Anisotropy (Owens [1974])	$A = (s_1 - s_3)/\chi_b$
Anisotropy Degree (Nagata [1961])	$P = \tau_1/\tau_3$
"Corrected" Anisotropy (Jelinek [1981])	$P' = \exp\sqrt{2[(\eta_1 - \bar{\eta})^2 + (\eta_2 - \bar{\eta})^2 + (\eta_3 - \bar{\eta})^2]}$
Shape:	
Shape Factor (Jelinek [1981])	$T = (2\eta_2 - \eta_1 - \eta_3)/(\eta_1 - \eta_3)$
Lineation (Balsley and Buddington [1960])	$L = \tau_1/\tau_2$
Foliation (Stacey et al. [1960])	$F = \tau_2/\tau_3$
log Lineation (Woodcock [1977])	$L' = \ln(L)$
log Foliation (Woodcock [1977])	$F' = \ln(F)$
Elongation (this book)	$E' = \tau_1 + .5\tau_3$
Roundness (this book)	$R = \sin(60)\tau_3$

5.10. Examples

• Example 5.1

Use the program **s_eigs** to calculate the eigenvalues and eigenvectors of the χ_i data in Table 5.2.

TABLE 5.2. Input data for Example 5.1

χ_{11}	χ_{22}	χ_{33}	χ_{12}	χ_{23}	χ_{13}
0.33412680	0.33282742	0.33304584	-.00015292	0.00124846	0.00135721
0.33556300	0.33198264	0.33245432	0.00087260	0.00024138	0.00096168
0.33584908	0.33140624	0.33274472	0.00131844	0.00118815	0.00002988
0.33479759	0.33142531	0.33377719	-.00047491	0.00049540	0.00044300
0.33505613	0.33114848	0.33379537	-.00101373	0.00028536	0.00034849
0.33406156	0.33226913	0.33366925	-.00002265	0.00098544	0.00005555
0.33486593	0.33216035	0.33297372	-.00035492	0.00039254	0.00015402
0.33510643	0.33196402	0.33292958	0.00075968	0.00057242	0.00010110

Solution

Put the data into a file called **ex5.1**. Then type the following:

% **s_eigs** < **ex5.1** > **ex5.1a**

and the computer responds by putting the eigenvalues and directions of the associated eigenvectors for each record as listed in Table 5.3 into **ex5.1a**. All eigenvectors are mapped to the lower hemisphere.

TABLE 5.3. Output data from Example 5.1.

τ_3	Dec.	Inc.	τ_2	Dec.	Inc.	τ_1	Dec.	In
0.33127224	238.60	44.99	0.33351420	127.32	19.96	0.33521362	20.66	38.2
0.33178045	282.36	2.54	0.33218098	183.25	74.32	0.33603854	13.06	15.4
0.33047004	283.62	26.77	0.33328291	118.33	62.45	0.33624708	16.67	6.0
0.33123816	260.84	12.43	0.33377632	138.38	67.68	0.33498562	355.01	18.2
0.33085684	255.86	7.02	0.33379161	131.50	77.69	0.33535152	347.11	10.0
0.33175950	268.72	27.32	0.33405055	171.02	14.54	0.33418989	56.02	58.4
0.33195050	261.85	21.86	0.33313221	90.81	67.90	0.33491729	353.10	3.1
0.33157604	281.50	21.78	0.33312118	117.51	67.43	0.33530281	13.76	5.6

● **Example 5.2**
It is often useful to go "backwards" from the eigenparameters to the matrix elements. Use **eigs_s** to convert the output from Example 5.1 back to matrix elements.
Solution
Type:
% **eigs_s< ex5.1a > ex5.2**
to get back the matrix elements (see Table 5.4).
Notice how these are slightly different from the original data, because of round-off error; hence the transformation between matrix elements and eigenvalues is rather unstable and should be done as little as possible.

TABLE 5.4. Output from Example 5.2.

χ_{11}	χ_{22}	χ_{33}	χ_{12}	χ_{23}	χ_{13}
0.33417818	0.33278573	0.33303612	-.00016533	0.00123392	0.00134539
0.33557811	0.33197480	0.33244714	0.00085172	0.00026042	0.00094949
0.33584759	0.33142728	0.33272523	0.00131540	0.00120533	0.00003255
0.33479103	0.33140954	0.33379945	-.00044067	0.00048470	0.00045752
0.33505136	0.33115500	0.33379373	-.00102342	0.00029002	0.00034980
0.33406496	0.33225048	0.33368459	-.00002767	0.00097086	0.00006390
0.33486152	0.33214658	0.33299193	-.00036709	0.00037305	0.00016616
0.33511242	0.33195129	0.33293635	0.00075242	0.00056089	0.00009528

● **Example 5.3**
Anisotropy data come from specimens with an arbitrary orientation. Use **s_geo** to rotate sets of six matrix elements referenced to an orientation arrow (along X_1) with given azimuth and plunge into geographic coordinates. Use the data in Table 5.5.
Solution
Put the data into a file **ex5.3a** and type:
% **s_geo < ex5.3a**
The computer responds with **ex5.1** from the first example. Now rotate the data in **ex5.1** using a strike of 204 and a dip of 25. First attach the desired strike and dip to the data by:
% **awk '{print $1,$2,$3,$4,$5,$6,204,25}' ex5.1 > ex5.3b**
Then type:
% **s_tilt < ex5.3b**
to get the data in Table 5.6.

TABLE 5.5. Input data for Example 5.3.

χ_{11}	χ_{22}	χ_{33}	χ_{12}	χ_{23}	χ_{13}	Az.	Pl.
0.331470	0.334140	0.334390	0.000751	-.000834	-.000167	80.00	-46.00
0.333359	0.333359	0.333281	-.001555	-.001322	0.001166	52.00	-23.00
0.330976	0.335736	0.333288	0.001632	0.000136	0.000000	123.00	-29.00
0.331500	0.334654	0.333846	-.000647	-.000566	-.000485	68.00	-26.00
0.331220	0.335220	0.333568	-.000470	-.000470	-.000861	68.00	-26.00
0.331796	0.334056	0.334148	-.000092	-.000046	-.000277	86.00	-34.00
0.332432	0.334399	0.333169	0.001066	0.000328	0.000000	108.00	-31.00
0.331755	0.335127	0.333118	0.000789	0.000000	-.000072	115.00	-25.00

TABLE 5.6. Data output of Example 5.3.

χ_{11}	χ_{22}	χ_{33}	χ_{12}	χ_{23}	χ_{13}
0.33455712	0.33192652	0.33351636	-.00043563	0.00092772	0.00105001
0.33585498	0.33191571	0.33222941	0.00055965	-.00005314	0.00064736
0.33586675	0.33084911	0.33328408	0.00142253	0.00013233	0.00009207
0.33488667	0.33138487	0.33372846	-.00056591	-.00039077	0.00004880
0.33506599	0.33127031	0.33366376	-.00105201	-.00057259	-.00029960
0.33407679	0.33177570	0.33414757	0.00007003	0.00018452	0.00005064
0.33483920	0.33197856	0.33318222	-.00028460	0.00003518	-.00029258
0.33513147	0.33175039	0.33311826	0.00077915	-.00006385	0.00004614

• **Example 5.4**

Use **k15_s** to calculate the best-fit tensor elements and residual error for data in Table 5.7. These are: the sample name, azimuth and plunge, and strike and dip, followed by the fifteen measurements made using the scheme in Figure 5.2. Calculate the s data in geographic and tilt adjusted coordinates.

Solution

Enter the data into datafile **ex5.4**. To calculate the matrix elements in specimen coordinates, type:

% **k15_s** < **ex5.4**

and the computer responds with the data in Table 5.8.

To do just the geographic rotation, type:

% **k15_s -g** < **ex5.4**

The computer should respond with the data as in Table 5.2, but with σ from the previous example at the end of each record.

TABLE 5.7. Input data for Example 5.4.

tr245f	80.00	-46.00 204 25		
995.	999.	993.	995.	1000.
1004.	999.	1001.	1004.	999.
998.	997.	1002.	998.	997.
tr245g	52.00	-23.00 204 25		
1076.	1066.	1072.	1077.	1067.
1076.	1068.	1072.	1076.	1067.
1068.	1075.	1071.	1068.	1076.
tr245h	123.00	-29.00 204 25		
1219.	1231.	1218.	1220.	1232.
1230.	1231.	1234.	1230.	1231.
1221.	1221.	1225.	1221.	1221.
tr245i1	68.00	-26.00 204 25		
1031.	1027.	1025.	1032.	1028.
1035.	1031.	1035.	1035.	1032.
1030.	1027.	1032.	1030.	1027.
tr245i2	68.00	-26.00 204 25		
1065.	1062.	1058.	1066.	1063.
1069.	1067.	1071.	1070.	1066.
1064.	1059.	1065.	1065.	1059.
tr245j	86.00	-34.00 204 25		
1804.	1803.	1799.	1805.	1804.
1811.	1810.	1811.	1811.	1811.
1806.	1803.	1811.	1806.	1803.
tr245k	108.00	-31.00 204 25		
1013.	1019.	1014.	1014.	1021.
1017.	1019.	1020.	1017.	1019.
1015.	1015.	1016.	1015.	1015.
tr245l	115.00	-25.00 204 25		
1159.	1164.	1156.	1159.	1165.
1164.	1164.	1168.	1164.	1164.
1158.	1158.	1161.	1159.	1158.

Finally, to do the geographic and tectonic rotations, type:

% k15_s -t < ex5.4

to get the data in Table 5.9.

TABLE 5.8. Output from Example 5.4.

χ_{11}	χ_{22}	χ_{33}	χ_{12}	χ_{23}	χ_{13}	σ
0.3314698	0.3341399	0.3343902	0.0007509	-.0008343	-.0001668	0.0000861
0.3333592	0.3333592	0.3332814	-.0015552	-.0013219	0.0011664	0.0001719
0.3309763	0.3357356	0.3332880	0.0016317	0.0001359	0.0000000	0.0001813
0.3315003	0.3346542	0.3338455	-.0006469	-.0005660	-.0004852	0.0001486
0.3312198	0.3352119	0.3335681	-.0004696	-.0004696	-.0008610	0.0001837
0.3317956	0.3340559	0.3341482	-.0000922	-.0000461	-.0002767	0.0001047
0.3324315	0.3343989	0.3331694	0.0010656	0.0003278	0.0000000	0.0001762
0.3317548	0.3351271	0.3331181	0.0007892	0.0000000	-.0000717	0.0001111

TABLE 5.9. Output from Example 5.4.

χ_{11}	χ_{22}	χ_{33}	χ_{12}	χ_{23}	χ_{13}	σ
0.3345570	0.3319266	0.3335163	-.0004357	0 0009277	0.0010500	0.0000861
0.3358549	0.3319157	0.3322294	0.0005597	-.0000531	0.0006474	0.0001719
0.3358667	0.3308492	0.3332841	0.0014225	0.0001322	0.0000922	0.0001813
0.3348866	0.3313848	0.3337284	-.0005658	-.0003908	0.0000487	0.0001486
0.3350661	0.3312701	0.3336637	-.0010517	-.0005725	-.0002996	0.0001837
0.3340768	0.3317756	0.3341475	0.0000700	0.0001844	0.0000507	0.0001047
0.3348392	0.3319785	0.3331822	-.0002845	0.0000351	-.0002926	0.0001762
0.3351314	0.3317503	0.3331182	0.0007791	-.0000637	0.0000461	0.0001111

• **Example 5.5**

Use **k15_hext** to calculate statistics for the data in **ex5.4** using the linear propogation assumptions. Calculate the following: bulk susceptibility, F, F12, F23, E12, E13, E23 and the assorted directions.

Repeat the same calculations for geographic and tilt adjusted coordinates.
Solution

For specimen coordinates, type
% **k15_hext** < **ex5.4**
for:
tr245f bulk susceptibility = 998.733
F = 418.84 F12 = 338.36 F23 = 194.46
0.33521 256.1 45.9 1.8 165.0 1.0 4.2 74.1 44.1
0.33351 74.1 44.1 3.2 165.0 1.0 4.2 256.1 45.9

0.33127 165.0 1.0 1.8 256.1 45.9 3.2 74.1 44.1
tr245g bulk susceptibility = 1071.67
F = 148.81 F12 = 2.71 F23 = 251.69
0.33604 314.0 32.6 3.4 51.9 12.0 3.7 159.4 54.8
0.33218 159.4 54.8 3.7 314.0 32.6 32.1 51.9 12.0
0.33178 51.9 12.0 3.4 314.0 32.6 32.1 159.4 54.8
tr245h bulk susceptibility = 1225.67
F = 202.61 F12 = 120.34 F23 = 133.64
0.33625 72.8 2.5 2.6 342.8 0.8 5.1 234.7 87.4
0.33328 234.7 87.4 5.1 72.8 2.5 5.4 342.8 0.8
0.33047 342.8 0.8 2.6 72.8 2.5 5.4 234.7 87.4

$$\vdots$$

For geographic coordinates, type
%k15_hext -g < ex5.4
and you will get:
  tr245f bulk susceptibility = 998.733
  F = 420.98 F12 = 338.40 F23 = 194.44
  0.33521 20.7 38.3 1.8 238.6 45.0 4.2 127.3 20.0
  0.33351 127.3 20.0 3.2 238.6 45.0 4.2 20.7 38.3
  0.33127 238.6 45.0 1.8 20.7 38.3 3.2 127.3 20.0
  tr245g bulk susceptibility = 1071.67
  F = 148.82 F12 = 2.71 F23 = 251.69
  0.33604 13.1 15.5 3.4 282.4 2.5 3.7 183.3 74.3
  0.33218 183.3 74.3 3.7 13.1 15.5 32.1 282.4 2.5
  0.33178 282.4 2.5 3.4 13.1 15.5 32.1 183.3 74.3
  tr245h bulk susceptibility = 1225.67
  F = 203.10 F12 = 120.35 F23 = 133.64
  0.33625 16.7 6.0 2.6 283.6 26.8 5.1 118.3 62.5
  0.33328 118.3 62.5 5.1 16.7 6.0 5.4 283.6 26.8
  0.33047 283.6 26.8 2.6 16.7 6.0 5.4 118.3 62.5

$$\vdots$$

For tilt adjusted coordinates, type
k15_hext -t < ex5.4
and you will get:
  tr245f bulk susceptibility = 998.733
  F = 419.92 F12 = 338.40 F23 = 194.44
  0.33521 2.8 32.8 1.8 252.7 28.1 4.2 131.5 44.1
  0.33351 131.5 44.1 3.2 252.7 28.1 4.2 2.8 32.8
  0.33127 252.7 28.1 1.8 2.8 32.8 3.2 131.5 44.1

tr245g bulk susceptibility = 1071.67
F = 149.35 F12 = 2.71 F23 = 251.69
0.33604 7.6 9.5 3.4 101.5 21.9 3.7 255.7 65.9
0.33218 255.7 65.9 3.7 7.6 9.5 32.1 101.5 21.9
0.33178 101.5 21.9 3.4 7.6 9.5 32.1 255.7 65.9
tr245h bulk susceptibility = 1225.67
F = 202.85 F12 = 120.34 F23 = 133.64
0.33625 14.8 2.4 2.6 284.7 2.1 5.1 153.0 86.8
0.33328 153.0 86.8 5.1 14.8 2.4 5.4 284.7 2.1
0.33047 284.7 2.1 2.6 14.8 2.4 5.4 153.0 86.8

$\vdots$

• **Example 5.6**
Use **k15_hext** to calculate the statistics of the whole file **ex5.4**. Repeat
the excercise for geographic and tilt adjusted coordinates.
Solution
Type:
% **k15_hext -a < ex5.4**
The **[-a]** switch tells the program to average over the whole file and the
response is:
F = 2.55 F12 = 2.16 F23 = 1.12
0.33471 265.8 17.6 19.6 356.5 2.1 40.4 93.0 72.2
0.33349 93.0 72.2 31.5 356.5 2.1 40.4 265.8 17.6
0.33180 356.5 2.1 19.6 265.8 17.6 31.5 93.0 72.2
For geographic coordinates, type
k15_hext -ag < ex5.4
and get:
% **k15_hext -ag < ex5.4**
F = 5.77 F12 = 3.66 F23 = 3.55
0.33505 5.3 14.7 13.3 268.8 23.6 25.5 124.5 61.7
0.33334 124.5 61.7 25.1 268.8 23.6 25.5 5.3 14.7
0.33161 268.8 23.6 13.3 5.3 14.7 25.1 124.5 61.7
For tilt adjustment, type
% **k15_hext -at < ex5.4**
and get:
F = 6.08 F12 = 3.86 F23 = 3.74
0.33505 1.1 5.7 13.0 271.0 0.7 24.9 173.8 84.3
0.33334 173.8 84.3 24.6 271.0 0.7 24.9 1.1 5.7
0.33161 271.0 0.7 13.0 1.1 5.7 24.6 173.8 84.3

● **Example 5.7**

Use **s_hext** to calculate statistics from the averaged matrix elements obtained from **k15_s** and **s_geo** in **ex5.4**.

Solution

The outcome of Example 5.4 was the input for Example 5.1, so

% **s_hext < ex5.1**

to get:

F = 5.77 F12 = 3.55 F23 = 3.66
N = 8 sigma = 6.41813E-04
0.33505 5.3 14.7 13.3 268.8 23.6 25.5 124.5 61.7
0.33334 124.5 61.7 25.1 268.8 23.6 25.5 5.3 14.7
0.33161 268.8 23.6 13.3 5.3 14.7 25.1 124.5 61

which you will notice is identical to the outcome of Example 5.6. Also, note that while barely anisotropic (F > 3.48), these data fail the discrimination tests F12 and F13, suggesting that in fact τ_1, τ_2, and τ_3 cannot be discriminated.

● **Example 5.8**

Calculate confidence ellipses of Jelinek [1978] for **ex5.1** using **s_jel78** for the data in **ex5.1**.

Solution

Type:

% **s_jel78 < ex5.1**

to get:

N = 8
0.33505 5.3 14.7 15.1 268.8 23.6 16.8 124.5 61.7
0.33334 124.5 61.7 15.8 268.8 23.6 16.8 5.3 14.7
0.33161 268.8 23.6 15.1 5.3 14.7 15.8 124.5 61.7

● **Example 5.9**

Calculate bootstrap statistics for the data in **ex5.4** (transformed into geographic coordinates) using **bootams**. Repeat using the parametric bootstrap option.

Solution

First the data in **ex5.4** must be converted to matrix elements, by the command **k15_s -g < ex5.4 > ex5.9** as in Example 5.4. Then, type:

% **bootams < ex5.9** to get:

0.33505 0.00021 5.3 14.7 11.6 264.4 35.8 14.2 113.8 50.4
0.33334 0.00021 124.5 61.7 7.2 224.4 5.3 18.3 317.1 27.7
0.33161 0.00016 268.8 23.6 12.3 10.3 24.6 13.6 140.7 54.7

For a parametric bootstrap, type:

bootams -p < ex5.9 to get:

0.33505 0.00021 5.3 14.7 11.1 263.4 38.3 14.1 112.3 48.0

0.33334 0.00021 124.5 61.7 6.0 225.6 5.9 18.1 318.7 27.5

0.33161 0.00014 268.8 23.6 11.9 2.3 8.1 12.9 110.1 64.9

These measurements have very low signal/noise ratios, hence the outcomes of the simple and parametric bootstraps are similar.

Also note that, according to the standard deviations of the bootstrapped τ values, the three eigenvalues are significantly different, in contrast to the F test results from Examples 5.6 and 5.7

• Example 5.10

1) Compare the four methods of calculating confidence ellipses (parametric and non-parametric bootstrap, Hext [1963], and Jelinek [1978]) using **plotams** on the data from Example 5.9. 2) Plot the distribution of eigenvectors obtained from the site parametric bootstrap in equal area projection.

Solution

1) Type:

% **plotams -pxj < ex5.9 | plotxy**

to get the postscript file **mypost** as shown in Figure 5.15. Compare these ellipses with the outcomes of **s_hext, s_jel78, bootams, bootams -p** in previous examples. The two bootstrap methods are quite similar, the parametric one being slightly fatter in the V_1 ellipse. The Jelinek [1978] ellipses are rounder than the bootstrapped ones. Those of Hext [1963] are much larger than the others.

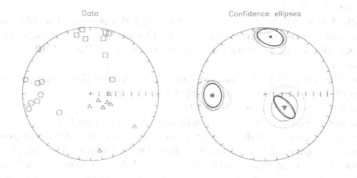

Figure 5.15. Equal area projection of data and confidence ellipses from Example 5.10, using the command: **plotams -pxj < ex5.9 | plotxy**.

2) Now type:

% **plotams -Pv** < **ex5.9** | **plotxy**
to get plot shown in Figure 5.16.

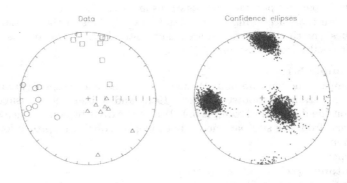

Figure 5.16.   Equal area projection of eigenvectors of the data in file **ex5.9** and the bootstrapped eigenvectors from Example 5.10. Plot produced by the command: **plotams -Pv** < **ex5.9** | **plotxy**.

● **Example 5.11**
Check if the anisotropy data from two chilled margins of a dike indicate imbrication using **plotams** and **s_hist**. The eastern margin data are those in Example 5.4 and the western margin data are in Table 5.10.
Solution
Enter the data into a file named **ex5.11a**. Perform the geographic corrections and specimen averaging using **k15_s** by typing:
% **k15_s -g** < **ex5.11a** > **ex5.11b**
(The data from Example 5.4 should already be in **ex5.9**.)
To examine these new data, use **plotams** as before to get the **mypost** plot shown in Figure 5.17. Compare this figure with Figure 5.15. Can the eigenvectors be distinguished on the basis of confidence ellipses alone? Another way to consider the problem is to compare histograms of the the mean eigenvectors of interest (in this case the principal one) generated during bootstrapping to see if any humps can be distinguished. **s_hist** will do this with the appropriate switches.
To see the options available, type:
% **s_hist -h**
or check Appendix 1.

TABLE 5.10. Western margin data for
Example 5.11.

tr245a	124.00	-26.00		
1521.	1525.	1515.	1520.	1526.
1518.	1526.	1534.	1518.	1527.
1511.	1513.	1512.	1512.	1513.
tr245b	44.00	-67.00		
1718.	1716.	1717.	1719.	1716.
1730.	1726.	1718.	1731.	1726.
1722.	1734.	1739.	1723.	1734.
tr245c	154.00	-45.00		
1668.	1685.	1671.	1669.	1685.
1673.	1684.	1682.	1673.	1685.
1670.	1677.	1675.	1670.	1676.
tr245d	86.00	-53.00		
1883.	1861.	1866.	1884.	1862.
1877.	1874.	1880.	1877.	1874.
1870.	1866.	1871.	1870.	1867.
tr245e	80.00	-54.00		
1732.	1712.	1716.	1732.	1712.
1728.	1720.	1728.	1728.	1720.
1717.	1719.	1720.	1717.	1719.

To select a parametric bootstrap, because we have the σ values already,
we choose -p. To compare data from two files, we select the -c option
and supply the two file names. To plot the principal eigenvector, we select
the -1 option and to plot the 95% confidence bounds for the respective
bootstrapped eigenparameters, we select the -b option. Thus, we type:
% s_hist -pcb1 ex5.9 ex5.11b | plotxy
and get the **mypost** file shown in Figure 5.18. The 95% confidence intervals
for the x_2 component do not overlap, hence the two data sets are discrete,
indicating imbrication and allowing interpretation of flow direction.

● **Example 5.12**
Use s_hist to see what average shape the data in **ex5.9** is. Test whether
$\tau_1 > \tau_2 > \tau_3$.
Solution
We want to plot the eigenvalues generated by the parametric bootstrap and
the 95% confidence intervals for each eigenvalue. To do this type:
% s_hist -ptb < ex5.9 | plotxy
and get Figure 5.19.

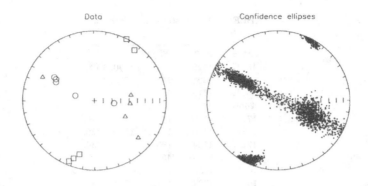

Figure 5.17. Equal area projection of eigenvectors from the data in file **ex5.11b** along with the bootstrapped eigenvectors from Example 5.11, using the command: **plotams -vP < ex5.11b | plotxy**.

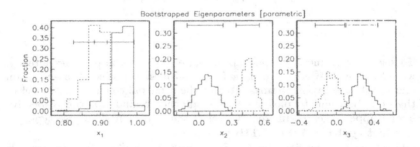

Figure 5.18. Histograms from Example 5.11, generated using the command: **s_hist -pcb1 ex5.9 ex5.11b | plotxy**.

• Example 5.13

Use **s_flinn** to plot Flinn and Ramsay diagrams using a parametric bootstrap of the data in **ex5.9**. Use **s_pt** to make a Jelinek diagram of the same data.

Solution

For a Flinn diagram (parametric bootstrap) as shown in Figure 5.20, type:

% **s_flinn -p < ex5.9 | plotxy**

For a Ramsay diagram (log Flinn) as shown in Figure 5.21, type:

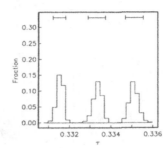

Figure 5.19. Histograms from Example 5.12, generated using the command: **s_hist -ptb < ex5.9 | plotxy**.

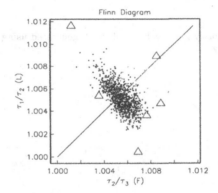

Figure 5.20. Flinn plot for Example 5.13, generated using the command: **s_flinn -p < ex5.9 | plotxy**.

% **s_flinn -pl < ex5.9 | plotxy**
For a Jelinek diagram as shown in Figure 5.22, type:
% **s_pt -p < ex5.9 | plotxy**

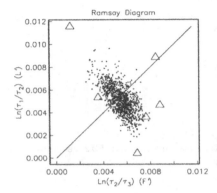

Figure 5.21. Ramsay plot for Example 5.13, generated using the command: **s_flinn -pl**
< ex5.9 | plotxy

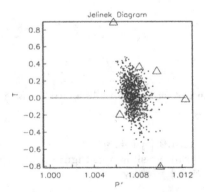

Figure 5.22. Jelinek plot for Example 5.13, generated using the command: **s_pt -p <**
ex5.9 | plotxy.

Chapter 6

PALEOMAGNETIC APPLICATIONS

In this book we have reviewed the basic tools of paleomagnetism. To first order, the geomagnetic field is dipolar with the axis of the field corresponding to the spin axis. We learned how rocks could retain a remanence that reflects the Earth's magnetic field. In subsequent chapters, we developed statistical tools for dealing with paleomagnetic data. In this chapter, we will briefly examine gelogical applications of these paleomagnetic tools.

6.1. Magnetostratigraphic applications

An important application of the fact that the geomagnetic field undergoes frequent reversals, whose ages are fairly well known, at least for the last hundred million years or so, is to use the reversal time scale as a dating tool for stratigraphic sequences. The pattern of polarity zones is determined by measuring the magnetization of samples taken from the stratigraphic section as shown in Figure 6.1. If the polarity zones in the so-called *magnetostratigraphy* can be unambiguously correlated to the GPTS, they constitute a precise temporal framework for sedimentary or volcanic sequences. Such records have proved invaluable for correlating stratigraphic information on a global basis and are the primary means for calibrating the Cenozoic fossil record with respect to time.

Sedimentation is not always a continuous process in many environments and a stratigraphic section may have gaps of significant duration. Also, the magnetic recording process of the rock may be unreliable over all or part of the section. Furthermore, incomplete sampling may give a polarity log that is aliased. For these reasons, there must be ways of establishing the reliability of a given polarity sequence and the robustness of a given correlation. For a more complete discussion of the subject of magnetostratigraphy, the reader is referred to the comprehensive book by Opdyke and Channell [1996] entitled *Magnetic Stratigraphy*. Briefly, the elements of a good magnetostratigraphic study include the following points.

- It must be established that a single component of magnetization can be (and has been) isolated by stepwise demagetization. To demonstrate this, examples of demagnetization data should be shown (see Chapter 3). There must also be a clear discussion of how directions were determined for each sample.
- Geological materials are not always perfect recorders of the geomagnetic field. It often happens that a given stratigraphic horizon has no consistent magnetization. Multiple samples per horizon (say three to five separately oriented samples) with coherent directions (i.e., non-random by tests such as those discussed in Chapters 3 and 4) indicate that the magnetization at a given level is reproducible. While it is not always possible to take multiple samples (for example from limited drill core

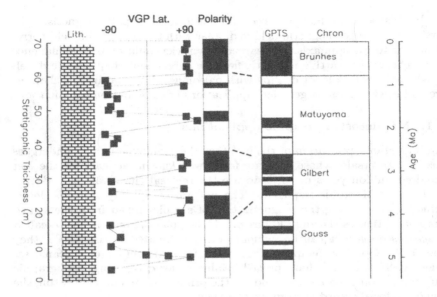

Figure 6.1. To the left is an idealized stratigraphic section. In the middle is a plot of hypothetical VGP latitude data derived from samples taken from the section (+90 is pointing north and -90 is pointing south). To the right is the interpreted polarity log. Correlation of the polarity log to the GPTS provides a means of dating the lithologic section.

material), it is alway desirable and certainly should be done whenever possible.

• The directional data must fall into two clearly separated groups, that are identifiable as either normal or reversed polarity. If fully oriented samples have been taken, the data can be plotted on an equal area projection (see Chapter 1) and/or subjected to the reversals test (Chapter 4). Often drill cores are not azimuthally oriented, and the paleomagnetic inclination is the only indicator of polarity. In this case, one can plot histograms of the vertical component and establish that the two polarities (positive and negative) have discrete "humps" at the values expected for the site latitude.

• The average direction should be compared with the reference field (the GAD field; see Chapter 1), and the expected direction based on the age of the formation for the sampling location. This can be done on an equal area projection, or in cartesian coordinates (using a bootstrap), as described in Chapter 4.

• Field tests (such as the fold test or conglomerate test as described in Chapter 3) that enable establishment of the age of magnetization relative to the rock formation are desirable.

• An independent estimate of the approximate age of the sequence is necessary. The better the age constraints, the more confident we can be in a given interpretation.

• The magnetostratigraphic pattern should match the polarity time scale. Few polarity zones should be ignored either in the section or in the time scale. Ideally, each polarity zone should be based on multiple sites in the section.

6.1.1. THE MAGNETOSTRATIGRAPHIC JACKKNIFE

In order to quantify the robustness of a given polarity sequence, Tauxe and Gallet [1991] defined a parameter J, that is sensitive to the dependence of a given polarity stratigraphy on the distribution of sampling sites. J is determined using a jackknife procedure. The jackknife is a statistical technique (see Efron [1982]) whereby a given parameter is calculated for a data set, and is then recalculated for the deleted data set with one or more data points. Repeated calculations for data sets with different data points deleted gives an idea of the variability of the parameter and the procedure can be used, among other things, for calculating uncertainty estimates.

To use the magnetostratigraphic jackknife (Figure 6.2), one counts the number of polarity reversals in a given section, then repeats the count after deletion of one or more sampling sites. In Figure 6.2a, we show the magnetostratigraphy obtained by Lowrie et al. [1982] from the Contessa Road section in the Umbrian Alps of Italy. The polarity pattern is based on 158 sampling sites and defines 19 polarity zones. The magnetostratigraphic jackknife is performed by random deletion of a given percentage of sampling sites followed by recounting the remaining polarity zones. This is repeated many times and the average number of polarity zones for a given number of sites (expressed as the percentage of the initial number of polarity zones) is plotted against the percentage of sites deleted (Figure 6.2b). The slope of the resulting line which relates the percentage of remaining polarity zones and percentage of sites deleted was defined as the magnetostratigraphic jackknife parameter J (Tauxe and Gallet [1991]). J for the Contessa Road magnetostratigraphy is -0.24. Based on numerous simulations, Tauxe and Gallet [1991] recommend J values in the range of 0 to -0.5 for a magnetostratigraphy to be robustly defined. The jackknife for the magnetostratigraphic pattern shown in Figure 6.1 is also shown in Figure 6.2b. The low value of J (-0.7) reflects the fact that the magnetostratigraphy is not well determined; several polarity zones are based on a single site. Addition or removal of single sites is therefore likely to result

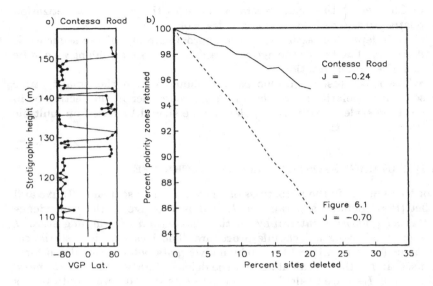

Figure 6.2. Illustration of the magnetostratigraphic jackknife: a) the magnetostratigraphic section from Contessa Road, Italy (Lowrie et al. [1982]) with 158 sampling sites and 19 polarity zones, b) the relationship between average percent of polarity zones retained and the percentage of sampling sites deleted. The slope of the line represents the magnetostratigraphic jackknife parameter J. Deletion of even 20% of the sampling sites from the Contessa Road stratigraphy shown in a) results in an average loss of only 4% of the polarity zones. Hence the magnetostratigraphic pattern is robust as reflected by the high J of -0.24. The magnetostratigraphy shown in Figure 6.1 is less robust with several polarity zones represented by single sampling sites. The low J value of -0.70 is an indication that the polarity pattern may be underdetermined.

in changes in the polarity pattern. In such a case, additional sampling is recommended.

6.1.2. TRACING OF MAGNETIC ISOCHRONS

Most magnetostratigraphic applications involve determination of a magnetostratigraphy through a stratigraphic sequence of sediments. Because polarity transitions occur relatively rapidly (in less than 4000 years, Chapter 1), the horizon bounding two polarity zones may represent an almost isochronous level. It is therefore possible to use magnetostratigraphy in a lateral sense, in order to delineate isochronous horizons within a given package of sediments (Behrensmeyer and Tauxe [1982]). In Figure 6.3, we show the application of magnetostratigraphy for tracing isochrons in a series of stratigraphic sections. The small sand body (darker gray) labeled "A" appears to have removed the normal polarity zone seen in sequences on the right of the figure either by erosion or because of unsuitable paleomagnetic properties of sand. Sand bodies B and C appear to represent quasi-isochronous horizons.

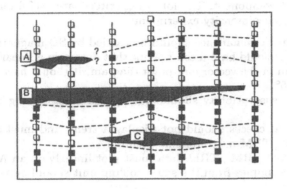

Figure 6.3. Application of magnetostratigraphic techniques for delineating isochronous horizons in a series of stratigraphic sections. The polarities of sampling sites are shown by open (reverse) and solid (normal) symbols. The light shading indicates silts, while the darker shaded units (labelled A-C) represent sand bodies, which were not suitable for paleomagnetic analysis in this example. The inferred isochrons (horizons that separate polarity zones) are shown as heavy dashed lines.

6.2. Paleointensity of the geomagnetic field

As discussed in Chapter 3, the geomagnetic field not only undergoes changes in direction but also changes in intensity. These intensity variations can be of great interest for several reasons. For example, the Earth's magnetic field partially shields the Earth from incoming cosmic rays (e.g., Elsasser et al. [1956]). Cosmic rays create radionuclides such as ^{14}C and ^{10}Be. Variations in the geomagnetic field intensity therefore result in variable production rates of these nuclides, an effect that must be taken into account if their decay is to be used for age determinations (see, e.g., Bard et al. [1990]). A second reason for interest in geomagnetic intensity variations is that, if these variations are accurately known, they could be used as a time scale that is potentially as effective as, for example, oxygen isotopic variations in deep sea sediments. Most importantly, geomagnetic field variations arise from processes deep in the Earth in the outer core. As such, they may hold a key to variations in boundary conditions of the outer core, or in convective styles. Also, because polarity reversals are always accompanied by low geomagnetic field intensities, perhaps intensity variations will provide some clues as to why and how the field reverses.

We discussed in Chapter 3 various methods for obtaining absolute paleointensity information from samples whose NRMs include TRM components and relative paleointensity information from samples whose NRMs include a DRM component. The following criteria are useful for judging a given absolute paleointensity experiment:

- The magnetic remanence should be carried by SD magnetite.
- The NRM should have a component that decays in a linear trajectory to the origin in the vector end-point diagram. It should have a MAD of less than 15°.
- Magnetic susceptibility should not change much during the experiment.
- The pTRM checks should not display a trend and must agree with the original pTRM measurement within 5%.
- The selected NRM-pTRM data must plot linearly on an Arai plot.
- Replicate samples from the same cooling unit should agree to within 10 - 15%.

The data shown in Figure 3.13 meet the highest standards, but the quality of data can vary substantially, as shown in Figure 6.4. In Figure 6.4a, we show data that provide a reasonable estimate of paleofield strength. The orthogonal plot of demagnetization data trends to the origin after removal of a soft component and the Arai diagram is well-behaved, with linear behavior and excellent pTRM checks. An iterative line search routine gives a best-fit line with a slope of -1.37. The laboratory field was 40 μT; hence the paleofield strength was 54 μT.

In Figure 6.4b, we show data that fail to meet minimum standards. The pTRM checks are poor, and the data are scattered. The data in the Arai plot are somewhat concave below the best-fit line. Data in Figure 6.4c are an example of unreliable data obtained from a Thellier-Thellier type experiment.

Relative paleointensity data have different reliability criteria (see King et al. [1983], Tauxe [1993], and Constable et al. [1998]). The following steps should be taken to insure the best possible relative paleointensity data:

- The natural remanence must be carried by stably magnetized magnetite, preferably in the grain size range of about 1-15 μm. The portion of the natural remanent vector used for paleointensity determination should be a single, well-defined component of magnetization. The nature of the remanent carrier can be assessed using stepwise demagneti-

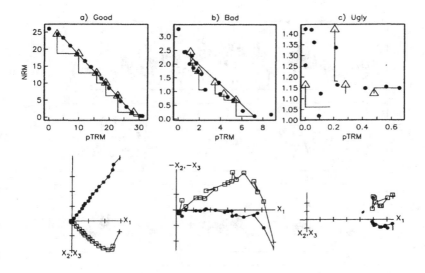

Figure 6.4. Representative results from a Thellier-Thellier type experiment. Top panel: Arai plots of pTRM intensity acquired versus NRM intensity remaining (dots) and pTRM checks (triangles). All intensities are in nAm^2 and the associated orthogonal plots are shown below. The samples were unoriented pieces of submarine basaltic glass. a) Specimen RD4011A from Pick and Tauxe [1993]. b) and c) T. Pick [unpublished].

zation combined with judicious use of various rock magnetic techniques, as described in Chapters 2 and 3.

• The detrital remanence must be an excellent recorder of the geomagnetic field (Tauxe [1993]). Normal and reversed polarity data should be antipodal (see Chapter 4). There is experimental evidence that the remanent intensity of unstirred sediments are not a simple linear function of field intensity. Thus, stirred (bioturbated) sediments are preferable to laminated sediments for relative paleointensity studies.

• Concentration of magnetic grains should not vary by of more than about an order of magnitude.

• Normalization should be done by several methods, all yielding consistent results. A portion of the Oligocene data from Hartl et al. [1993] are shown as an example in Figure 6.5. Correlation of the entire magnetostratigraphic pattern (Tauxe et al. [1983]) to the time scale of Cande and Kent [1995] suggests a C12n-C13n age range, as shown. Parameters

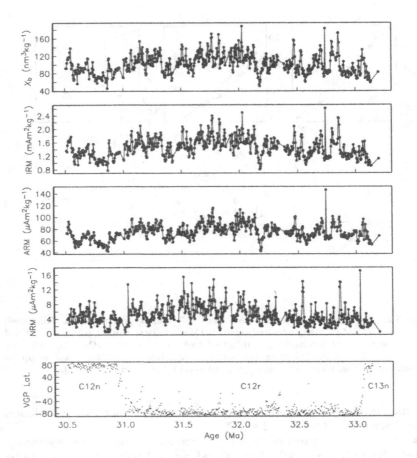

Figure 6.5.   Data from Hartl et al. [1993]. VGP Latitudes are calculated from the
NRM directions (after AF demagnetization to 17.5 mT). The pattern of VGP latitudes
was correlated to the GPTS chrons C12n, C12r and C13n by Tauxe et al. [1983]. The
time-scale of Cande and Kent [1995] gives ages as shown. NRM, ARM, IRM and χ_b data
are all normalized by sample mass.

such as χ_b, ARM or IRM can be plotted with respect to one another,
in a so-called *Banerjee plot* (Banerjee et al. [1981]). The various bulk
parameters should be linearly related to one another with a high correlation coefficient R. In the Oligocene data set, there is a strong degree

of correlation among the bulk parameters (Figure 6.6a,c) as reflected by
R values greater than 0.9.

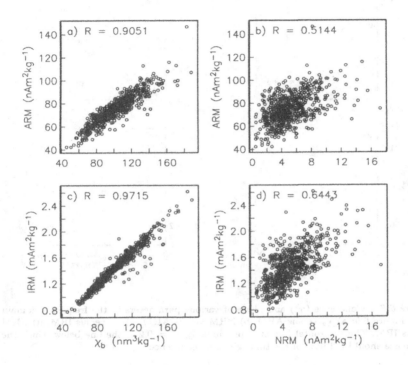

Figure 6.6. Correlation of various parameters from Figure 6.5: a) χ_b versus ARM, b)
NRM versus ARM, c) χ_b versus IRM, d) NRM versus IRM. R is the correlation coefficient.

Alternatively, Constable et al. [1998] pointed out the advantages of examining coherence γ^2 (the frequency domain analog of the correlation
coefficient) between the various bulk parameters (Figure 6.7). Analyzing data in the frequency domain allows us to examine the time-scales
over which the parameters behave coherently. As expected from the high
correlation coefficients, the coherence of IRM, ARM and χ_b is high, with
the most coherent being IRM and χ_b. Frequencies greater than about
100 Myr^{-1} indicate a drop in coherence, which reflects the increasing
dominance of noise at about the sampling interval (in this case, approximately 4 kyr).

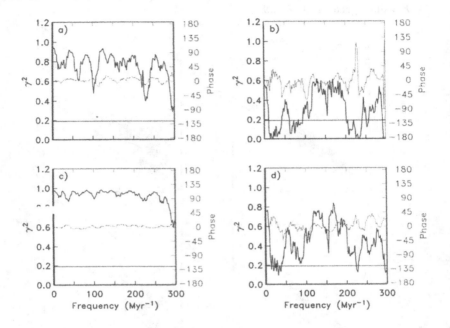

Figure 6.7. Coherence (γ^2) and phase of various parameters in the frequency domain from Figure 6.5: a) χ_b versus ARM, b) NRM versus ARM, c) χ_b versus IRM, d) NRM versus IRM. The horizontal line at approximately $\gamma^2 = 0.2$ is the line below which the coherence should fall 95% of the time, given no coherence.

• The estimated relative paleofield B^*, obtained by normalizing the NRM by some bulk parameter, can be re-normalized by the mean of the entire time series, such that the average relative paleointensity is unity (Figure 6.8). This allows comparison of different records with different average values. The agreement (or lack thereof) among the various estimates can be used as an indication of the reliability of the record. The normalizer of choice should be the one that is most correlated with the remanence (Constable et al. [1998]). As shown in Figures 6.6 and 6.7, IRM is the most highly correlated with NRM in the example data set, hence it is the best normalizer.

• Further confidence in relative paleointensity studies in sediments can be achieved by the use of Thellier-Thellier and pseudo-Thellier type experiments (Tauxe et al. [1995]).

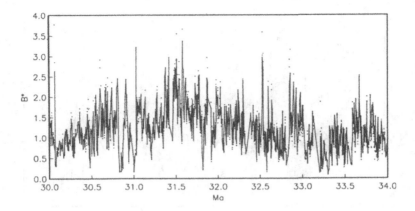

Figure 6.8. Relative paleointensity B^* for the data of Hartl et al. [1993], estimated by normalizing the NRM by IRM, ARM, and χ_b (dots). Each series is re-normalized by its mean. The solid line is through the IRM normalized data and is the best estimate of relative paleointensity for these data.

● Normalization by a suitable bulk magnetic parameter should significantly reduce the coherence of the relative paleointensity estimate with the normalizer and should increase coherence among different records from a given region; these should agree well within the limits of a common time-scale.

Recent analyses of many records distributed over the globe and which span the last 200,000 years, provide grounds or optimism that it may be possible to develop an independent time-scale based on relative paleointensity of the geomagnetic field. Guyodo and Valet [1996] showed that when placed on a common time-scale and scaled to a mean of unity, data sets from many disparate locations with different lithologies show similar trends (see Figure 6.9) which strongly suggests that the records have retained a substantial amount of paleofield information.

6.3. Paleomagnetic poles and apparent polar wander

As rock formations move, either as a result of local, regional or continental scale tectonic activity, they can carry with them a record of the direction of past magnetic fields. These records can be, as discussed in Chapter 1, approximately related to the position of the Earth's spin axis at the time

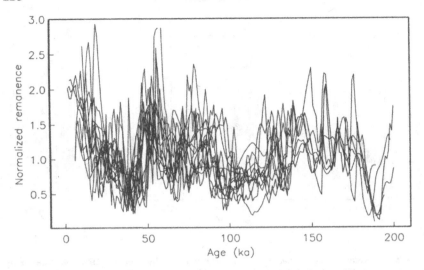

Figure 6.9. Compilation of Guyodo and Valet [1996] of relative paleointensity data from many locations around the world.

the rocks became magnetized. Thus, there are many tectonic applications for paleomagnetic data.

As continents move with respect to the North pole, the direction of the Earth's magnetic field changes when viewed from a single place. In the continent's frame of reference, the position of the magnetic pole appears to move. Tracks of these past pole positions are called *apparent polar wander paths* (APWPs). An example of a hypothetical APWP is shown in Figure 6.10.

There are several applications for paleomagnetic pole data, including constructing APWPs for single continents, constraining motions of tectonostratigraphic terranes, testing plate tectonic reconstructions for multiple continents, and, finally, testing the hypothesis that the paleomagnetic poles are consistent with some deep mantle reference frame. Tectonic applications of paleomagnetism have been thoroughly reviewed by many authors, from McElhinny [1973] to van der Voo [1993] and the reader is referred to these excellent books for an in-depth discussion. Briefly, the ingredients of reliable paleo-poles are as follows:

- The age of the formation must be known rather accurately.
- In order to average errors in orientation of the samples and scatter caused by secular variation, there must be a sufficient number of individually oriented samples from enough sites. What constitutes "sufficient"

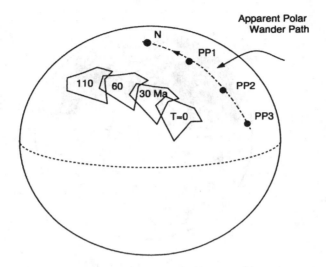

Figure 6.10. Illustration of the concept of apparent polar wander. If a continent was at the various locations labelled T = 0 to 110 Ma, paleomagnetic directions from those time intervals would correspond to paleopoles PP1, PP2, etc. in the present coordinate system. The poles appear to "wander".

and "enough" here is somewhat subjective and a matter of debate. Van der Voo [1990, 1993] recommends a minimum of 24 discrete samples of the geomagnetic field having a $\kappa > 10$. Some authors also compare scatter within and between sites in order to assess whether secular variation has been sufficiently sampled, but this relies on many assumptions as to what the magnitude of secular variation was, which is rather poorly known at present.

• It must be demonstrated that a coherent characteristic remanence component has been isolated by the demagnetization procedure.

• The age of the magnetization relative to the age of the rock should be addressed using field tests.

• There should be agreement in the pole position from units of similar age from a broad region.

• Be suspicious if a particular pole position falls on a younger part of the pole path or on the present field direction.

Consider the paleomagnetic poles plotted in Figure 6.11. These are all the poles (regardless of quality) compiled by van der Voo [1990] for cratonic North America. The poles form a smear that extends along arcs from the geographic North pole to south of the equator. The polarity of an ancient

Figure 6.11. Paleomagnetic poles from cratonic North America, as compiled by van der Voo [1990].

pole can be ambiguous, without a densely sampled track. The oldest poles on the diagram are from the early Cambrian. Given that rock units of increasingly older age become increasingly rare and that the paleomagnetic behavior becomes more uncertain to interpret with age, it can be difficult to do paleomagnetism in the Pre-Cambrian.

Data that meet minimum standards (three or more of the criteria described above) were grouped by age by van der Voo [1990]. These are plotted as circles in Figure 6.12. Also shown as triangles is a small selection of so-called *discordant poles* (van der Voo [1981]). What is immediately obvious is that the discordant poles do not fall anywhere near the APWP. Most are from western North America and indicate some clockwise rotations (the poles are rotated to the right of the expected poles). When taking into account the age of the formations, many also seem to have directions that are too shallow, which suggests possible northward transport of 1000's of kilometers. The validity and meaning of these discordant directions is still under debate, but it is obvious that most of the western Cordillera is not *in situ.*

One of the first uses of paleomagnetic data was as a test of the idea of *continental drift* (e.g., Wegener [1915]). Data from one continent, like those shown above, could be interpreted to indicate either motion of the continent with respect to a fixed geomagnetic pole, or motion of the geomagnetic pole

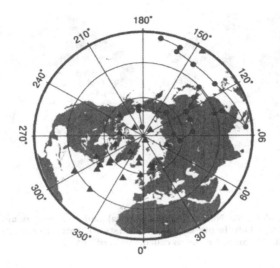

Figure 6.12. Circles are "reliable" poles from cratonic North America. (Data from van der Voo [1990]). So-called "discordant poles" from western North America are plotted as triangles (data from van der Voo [1981]).

with respect to a fixed continent. To test the hypothesis of continental drift, data from at least two continents are required.

In Figure 6.13, we plot the data compiled by van der Voo [1990], which meet the minimum standards of reliability, for North America and Europe. On the left, the poles are plotted with respect to present-day coordinates: the poles clearly fall on two separate tracks. This indicates that either the field was not at all dipolar, or that the two continents have moved, not only with respect to the geomagnetic pole, but also with respect to each other.

Many people who have contemplated the globe have had the desire to fit North and South America against Europe and Africa by closing the Atlantic Ocean. One such attempt, known as the *Bullard fit* (Bullard et al. [1965]), fits the continents together using misfit of a particular contour on the continental shelves as the primary criterion. Following van der Voo [1990], we rotate the European poles, using the Bullard fit, into North American coordinates on the right-hand side of Figure 6.13. After closing the Atlantic, the curves overlap rather well, and, if the ages of the poles are also taken into account, they also match well. The agreement provides strong support of the continental drift hypothesis and also of the Bullard fit.

In some studies, the focus is on discordant poles (e.g., Kamerling and

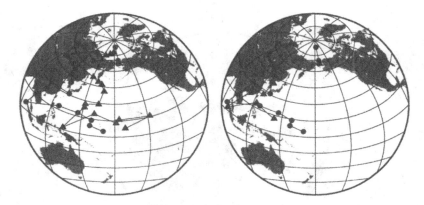

Figure 6.13.   Poles from North America (circles) and Europe (triangles). (Data from van der Voo [1990].) Left: in present day coordinates. Right: After rotation of Europe to close the Atlantic Ocean using the so-called "Bullard" fit.

Luyendyk [1979]). Regions with paleomagnetic directions that are significantly different from the direction expected from the appropriate reference pole of the APWP may have rotated or translated from their original positions as an independent entity (a tectonostratigraphic terrane of *microplate*). Paleomagnetic data for such studies should undergo the same scrutiny as for pole positions.

6.4. Magnetic fabric studies using AMS

Although much of paleomagnetism is devoted to the study of the magnetic vectors recorded in rocks, many areas of research seek to understand a more complex property of the magnetism of rocks, the behavior of rock magnetic tensors (see e.g., Tarling and Hrouda [1993]). Magnetic vectors yield information about past magnetic fields, whereas anisotropy in magnetic tensors yield information about the statistical alignment of magnetic crystals within the rock. Analysis of magnetic tensors has potential applications for unraveling the strain history, the fluid flow field at the rock's birth, etc. We will briefly outline several applications in the following.

6.4.1. PALEOCURRENT DIRECTIONS

Some of the earliest magnetic measurements made on sediments were of the anisotropy of magnetic susceptibility (see Ising [1942], Granar [1958], Rees [1965], Rees and Woodall [1975]). These studies and others (see summary

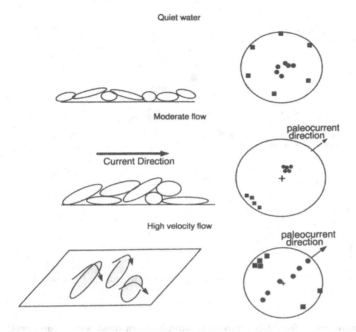

Figure 6.14. Characteristics of AMS data from sediments deposited in quiet water (top), moderate water flow (middle), and flow that is sufficient to entrain particles (bottom).

by Tarling and Hrouda [1993]) show that the magnetic fabric of sediments is strongly affected by the depositional environment (see Figure 6.14). We list the characteristics of the AMS ellipsoids generated in each type of sedimentary environment as follows:

• In quiet water conditions (see Figure 6.15):

    1) V_3 directions are perpendicular to the bedding plane, and
    2) the fabric is characterized by an oblate AMS ellipsoid.

• In moderate currents (no particle entrainment) (see Figure 6.15):

    1) particles are imbricated which results in (slightly) off-vertical V_3 directions,
    2) the V_1 direction (in lower hemisphere projections) is antiparallel to the paleo-flow direction, and
    3) the fabric is characterized by an oblate AMS ellipsoid.

• When deposition occurs on an inclined bedding plane:

    1) V_3 has a (slightly) off-vertical orientation,

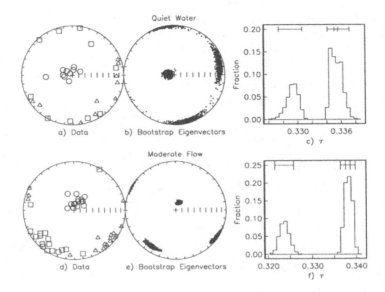

Figure 6.15. Examples of the AMS behavior in samples from quiet water (top panel) and moderate flow (bottom panel) regimes: a) and d) are lower hemisphere equal area projections of V_1 (squares), V_2 (triangles), and V_3 (circles) in tilt adjusted coordinates, b) and e) are the associated parametric bootstrap eigenvectors, c) and f) are histograms of the bootstrapped τs with associated 95% confidence intervals. Quiet water deposition has vertical V_3s and an oblate shape. Deposition in moderate flow has slightly off-vertical V_3, and an oblate shape. (The data courtesy of L. Masoni [unpublished].)

2) the V_1 direction (in lower hemisphere projections) is antiparallel to the direction of tilting, and

3) the fabric is characterized by an oblated AMS ellipsoid.

• When deposition occurs under high current flow (with particles entrained) (see Figure 6.16):

1) the V_3 distribution is streaked,

2) V_1 is perpendicular to the flow direction, and

3) the fabric is characterized by prolate or triaxial AMS ellipsoids.

• When tectonic disturbance controls the fabric (see Figure 6.16):

1) V_3 is parallel to the shortening direction (not necessarily perpendicular to bedding),

2) V_1 is parallel to the elongation direction, and

3) the fabric is characterized by prolate or triaxial AMS ellipsoids.

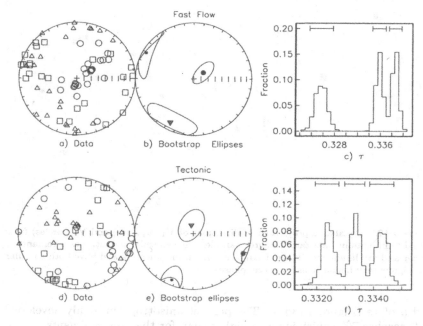

Figure 6.16. Examples of the AMS behavior in samples from fast flow (top) and tectonically affected (bottom) regimes: a) and d) are lower hemisphere equal area projections of V_1 (squares), V_2 (triangles), and V_3 (circles) in tilt-adjusted coordinates, b) and e) are the associated parametric bootstrap ellipses, c) and f) are histograms of the bootstrapped τs with associated 95% confidence intervals. Deposition in fast flow results in smearing of the V_3s and a triaxial shape. Tectonic disturbance results in deflection of V_3 into the plane of shortening, and a triaxial to prolate shape (data courtesy of L. Mason [unpublished]).

6.4.2. PALEOSOL DEVELOPMENT

Tauxe et al. [1990] explored how AMS data could be used to investigate the effects of paleosol formation on magnetic fabric. They showed that grey laminated sediments from fluvial deposits in Pakistan had specular hematite as the dominant magnetic phase and red paleosols were dominated by pigmentary plus specular hematite. In Figure 6.17, it appears that the grey sediments have nearly vertical V_3 directions, while the red sediments have no preferred orientation. In Figure 6.18 we plot the histograms of the bootstrapped τs, as discussed in Chapter 5. The grey sediments have dominantly oblate ellipsoid shapes (Figure 6.18a-c) while the red sediments exhibit a wide range of fabrics including oblate, spherical, triaxial

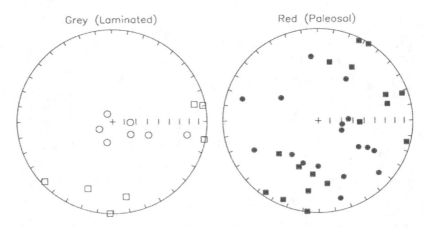

Figure 6.17. Equal-area projections of directions of V_1 (squares) and V_3 (circles) from fluvial sediments in Pakistan. The grey samples have specular hematite as a remanence carrier and are laminated. The red samples come from paleosols and have both specular and pigmentary hematite as magnetic phases.

and prolate (Figure 6.18d-h). The percent anisotropy in highly developed soils reaches 7%, which is much higher than for the grey sediments.

Representative histograms for the red and grey units, as shown in Figure 6.18, illustrate the range of observed AMS behavior, but it is difficult to show data for more than a few selected samples. We therefore need another way of examining the behavior of many samples at once. In Chapter 5, we discussed Flinn, Ramsay, Jelinek, and ternary diagrams for this purpose. The first three types of diagrams are useful for depicting oblate versus prolate shapes and display some measure of the degree of anisotropy. Because the range in ellipsoid shape for the paleosol study does not involve just oblate versus prolate shapes, but also spherical and triaxial shapes, we prefer the ternary diagram as being better suited for illustrating the full range of shapes.

In order to characterize the expected "primary" AMS ellipsoid, it is useful to examine hematite-bearing sediments that were redeposited in the laboratory. Lovlie and Torsvik [1984] and Stokking and Tauxe [1990] published relevant anisotropy data for such material. Data for both the depositional and the chemically precipitated fabrics are plotted in Figure 6.19. Also shown is the expected change in the values of the normalized eigenvalues as the sediment undergoes progressive alteration from pristine grey to heavily altered red. We expect a depositional fabric to plot near the data of Lovlie and Torsvik [1984]. Randomization of detrital grains would cause

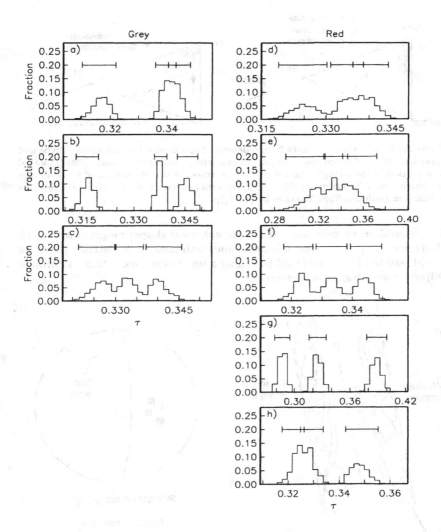

Figure 6.18.   Histograms of bootstrapped τs from representative grey and red samples. The grey samples tend to have only two distinct eigenvalues, with oblate shapes (a), but can range to triaxial (c). The red samples range from oblate (d), through spherical (e), and triaxial (f and g) to prolate (h).

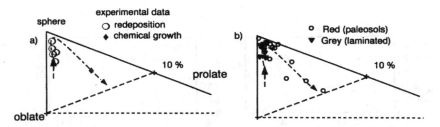

Figure 6.19. a) Anisotropy data from redeposited specularite-bearing sands (Lovlie and Torsvik [1984]) and chemically precipitated hematite (Stokking and Tauxe [1990]) plotted on a portion of the triangle. Also shown is the expected transformation of fabric during soil formation. b) Sedimentary AMS data from red and grey sediments from the Siwaliks in Pakistan (redrawn from Tauxe et al. [1990]).

the eigenvalues to move toward a more spherical shape; progressive growth of chemically precipitated hematite and realignment of detrital hematite would lead to eigenvalues that fall along a line toward more triaxial/prolate ellipsoids with higher anisotropies.

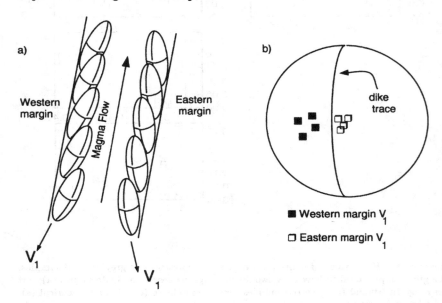

Figure 6.20. Principles of AMS for interpretation of flow directions in dikes (see Knight and Walker [1988]).

6.4.3. FLOW DIRECTIONS IN VOLCANIC DIKES

The principles by which flow directions can be determined in volcanic dikes were laid out by Knight and Walker [1988]. While the magma is flowing in the dike, elongate particles become imbricated against the chilled margins (see Figure 6.20). Opaque phases such as magnetite are often observed to be distributed along the fabric of the silicate phases (see Hargraves [1991]). The principal eigenvectors arising from such a *distribution anisotropy* parallel the fabric of the silicates. In Figure 6.20b, we show that in the ideal case, the V_1 directions from the two margins are distinct and fall on either side of the dike trace. Because the convention is to plot AMS data in lower hemisphere projections, the fact that the western margin data plot on the western side, and the eastern margin data plot on the eastern side suggests that the flow was upward. Thus, the AMS data from chilled margins of dikes can give not only a lineation, but a well constrained direction of magma flow. Following Tauxe et al. [1998], we recommend the following criteria for obtaining reliable flow direction data.

- Samples should be taken from within 10 cm of the chilled margin. This is necessary because the AMS of samples from the middle of dikes often has little to do with the direction of flow (Staudigel et al. [1992]).
- A minimum of six samples should be taken from each margin. Dikes can yield highly scattered results; a sufficient number of samples is necessary to establish the reliability of data for a given dike. Monte Carlo simulations by Tauxe et al. [1998] indicate that fewer than about five or six samples will yield nominal 95% confidence intervals that are too small.
- The bootstrapped eigenvalues for each margin must indicate a distinct τ_1 (either prolate or triaxial, as shown in Chapter 5).
- The bootstrapped v_{13} components (Figure 6.21) provide 95% confidence bounds on the inclination of V_1. If these exceed 30°, the data do not provide useful flow information.
- The mean V_1s should be within 45° of the dike orientation (v_{12} in dike coordinates should not exceed 0.707). If the principal directions are more than about 45° away from the plane of the dike, as in Figure 6.22a, the fabric is "inverse" and probably has nothing to do with the flow direction.
- Principal directions that differ in inclination by more than 30° from each other are termed "scissored" (Figure 6.22b). Scissoring can be caused by tectonism or can reflect a genuinely complicated flow pattern. Such data provide poor constraints for average flow directions in dikes.

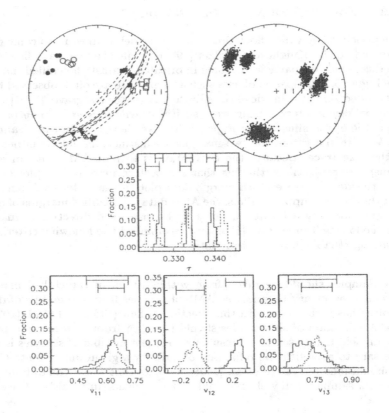

Figure 6.21. Top: equal area projections of AMS behavior for dike samples (conventions are the same as Figure 6.16). The left-hand equal area projection contains the eigenvector directions from the samples and the right-hand plot contains eigenvectors from bootstrapped para-data sets. Below the equal area projections are histograms of the bootstrapped eigenvalues (τ) and the cartesian coordinates of the principal eigenvectors (v_{1j}) after rotation into dike coordinates. The dike plane is shown by the dashed line in the v_{12} plot.

• The dike trace itself must be reasonably well constrained. If the dike is not quasi-planar, it is difficult to determine the relationship of the AMS data to the dike.

• The V_1 directions that pass the above criteria fall into three possible categories based on the behavior of the bootstrapped v_{12} components

(Figure 6.21). If the principal directions fall on opposite sides of the dike and are distinct from the dike plane (i.e., the confidence bounds of v_{12} do not overlap the trace of the dike plane as in the example), the data are classified 2D-A. If the confidence bounds from one margin overlap the dike plane, then the data fall in class 2D-B. If the data from neither margin are distinct from the dike plane, then the data fall in 2L. If only one margin yielded usable data, it is "1D" if the bootstrapped v_{12} components are distinct from the dike plane, otherwise it is "1L". Also, directions can be interpreted from "D" type classifications, while only lineations can be obtained from "L" type dikes.

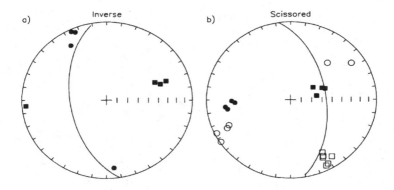

Figure 6.22.   a) Eigenvectors from a dike that shows an "inverse" fabric, with $\mathbf{V}_1$ nearly perpendicular to the dike plane (squares are $\mathbf{V}_1$, circles are $\mathbf{V}_3$, lower hemisphere projections). b) Eigenvectors for a dike exhibiting "scissored" directions, where the eastern (open symbols) and western (closed symbols) margins have virtually orthogonal $\mathbf{V}_1$ directions (redrawn from Tauxe et al. [1998]).

6.4.4. FLOW DIRECTIONS IN LAVA FLOWS

Buoyed by the success of using AMS for determining flow directions in dikes, several investigators have tried a similar method for determining flow directions in lava flows (see e.g., Cañon-Tapia et al. [1994]). Application to lava flows is much more difficult than for dikes. First, the flow direction is less well constrained. Lava flows tend to spill out with poorly controlled flow vectors. Examination of any one spot would yield a poor estimate of the average flow direction. Moreover, there are not two imbricated margins

to provide directional information, hence only a flow axis can be readily
interpreted.

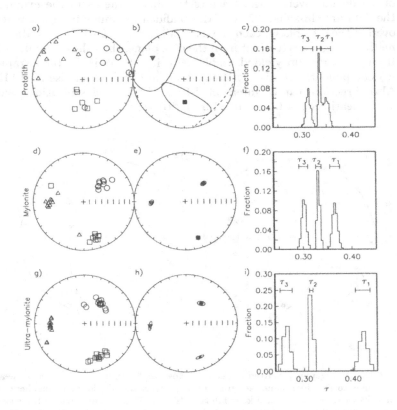

Figure 6.23. AMS data for increasing mylonitization. a) Eigenvectors from data obtained
from samples of the protolith plotted in equal area projection. V_1 are squares, V_2 are
triangles, and V_3 are circles. b) Approximate 95% confidence ellipses derived from the
bootstrap procedure. c) Histogram of bootstrapped eigenvalues (τ). d-f) Same as a-c)
but for mylonite samples. g-h) Same as a-c) but for ultra-mylonite samples. The data are
replotted from Housen et al. [1995].

6.4.5. MAGNETIC FABRIC IN METAMORPHIC ROCKS

To illustrate applications of AMS in metamorphic rocks, we choose the data
set of Housen et al. [1995] for Grenville metamorphic rocks (Figure 6.23).

These data come from three metamorphic grades: the relatively unde-
formed so-called "protolith", the "mylonite" and the highly strained "ultra-
mylonite". The eigenvectors of the protolith facies are poorly grouped and
the eigenvalues indicate an oblate shape. With increasing mylonitization,
the eigenvectors become more tightly clustered, with V_1 parallel to the regional lineation: the eigenvalue histograms indicate higher degrees of anisotropy and a more prolate shape (although the eigenvalues are still triaxial).

The histogram plots, when placed on the same horizontal scale, do give
a sense of degree of anisotropy and overall shape. Nonetheless, there are
other popular ways of depicting these characteristics, such as the Flinn,
Ramsay, and Jelinek diagrams and the Ternary diagram (see Chapter 5).

In Figures 6.24 and 6.25 we show the data of Figure 6.23 replotted on
Flinn, Ramsay, Jelinek, and Ternary diagrams. The results of progressive
deformation are evident in each plot and there is no reason to choose one
over the other. Thus, depending on the application, in contrast to the opin-
ion expressed by Tarling and Hrouda [1993], it is a matter of taste as to
which diagram is used.

6.5. Summary

In this book, we have reviewed the nuts and bolts of paleomagnetism. We
have discussed the tools of the trade: how to get samples, measure, and
analyze them. We have briefly considered some of the more popular applications. It is now up to the reader to use them. When thoughtfully applied,
the tools provided here can unleash the power of paleomagnetism.

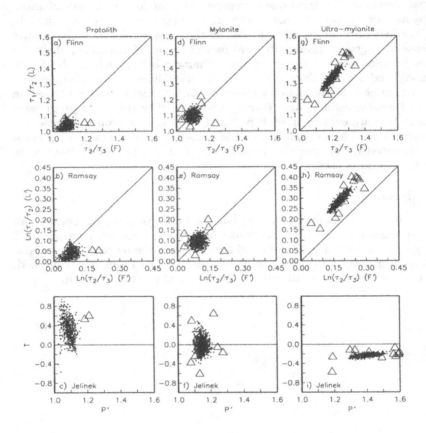

Figure 6.24. Data of Housen et al. [1995] plotted as Flinn, Ramsay, and Jelinek diagrams.

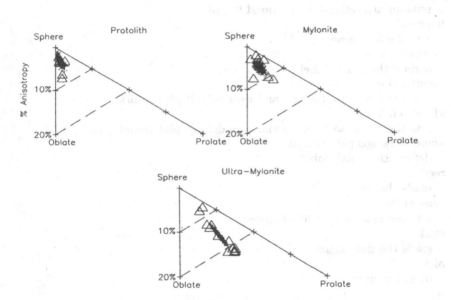

Figure 6.25.   Data of Housen et al. [1995] plotted as Ternary diagram.

6.6. Examples

• **Example 6.1**

Plot the magnetostratigraphic data (stratigraphic height, VGP latitude)
contained in file **ex6.1** using **plotxy**. Use the program **jackstrat** to investigate the robustness of the polarity pattern and to calculate J.

Solution

Type the following to put the plotting commands into a file **ex6.1.com**:

% **cat** > **ex6.1.com**

frame

   puts an attractive frame around the plot

file ex6.1

   sets the file name

ylabel VGP lat.

   defines the y axis label

ylimit 2 0 0

   sets the y axis to 2 inches and uses default plot limits

xlimit 6 0 0

   sets the x axis to 6 inches and uses default plot limits

xlabel Stratigraphic Height

   defines the x axis label

read

   reads the data

symbol 19

   sets the symbol to a filled circle

read

   reads the data again

plot 1 2

   plots the data

stop

   quits gracefully and writes the mypost file

Now type:

% **plotxy** < **ex6.1.com**

and see Figure 6.26.

The program **jackstrat** expects polarity information (positive or negative
numbers) in stratigraphic order. For this, you can use either the inclinations,
if they are steep enough to uniquely determine polarity, or, preferably, VGP
latitudes. In order to extract the polarity zonation from the file, we need
to select the second column (the VGP latitudes) from the file **ex6.1** using
the **awk** command. These can then be piped directly to **jackstrat**, whose
output in turn can be piped directly to **plotxy**:

% **awk** '{print $2}' ex6.1 | **jackstrat** | **plotxy**

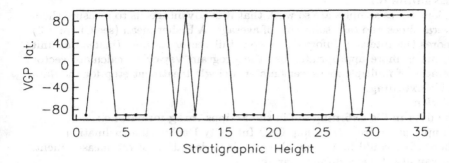

Figure 6.26. Output from Example 6.1, generated using the command **plotxy <
ex6.1.com**.

The output of the magnetostratigraphic jackknife is shown in Figure 6.27.

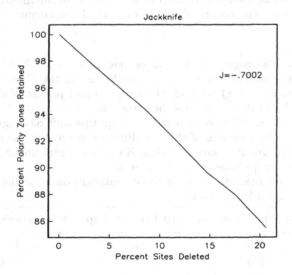

Figure 6.27. Output from Example 6.1, generated using the command: **awk '{print
$2}' ex6.1 | jackstrat | plotxy**. The *J* parameter (-0.7) is quite low, which reflects the
many polarity zones based on single samples (Figure 6.26).

• Example 6.2

1) Sometimes samples are so weak that it is advantageous to measure them several times and take some sort of average. A Fisher mean (see Chapter 3) ignores the intensity information, so a full vector average (instead of unit vector) is more appropriate. Use the program **vspec** to calculate vector averages of multiple measurements at a single treatment step for the data in file **ex6.mag**.

Solution

The data in file **ex6.mag** are in the Scripps **.mag** format, i.e.:

 Sample_name treatment_step CSD Intensity Declination Inclination

where CSD could be the circular standard deviation of the measurement, but can also be any dummy variable.

Now type:

% **vspec** < **ex6.mag** > **ex6.vmag**

to get the averaged data in **ex6.vmag**.

Unique sample/treatment steps are simply copied over and the vector average of multiple measurement replaces those data. The CSD field is replaced with a parameter similar to Fisher's R parameter, but takes into account the intensity. It is the resultant divided by the sum of all the intensities and ranges from zero to one, one resulting from perfect alignment of multiple measurements.

• Example 6.3

It is handy to have a program to help enter orientation data. In Chapter 3, we discussed three methods of orienting samples, using the "hand sample" (or "cube" in the following) method, the azimuth and plunge of the direction of drill (or "drill direction") method and the "sun compass" method. There are many more. What is required by programs such as **di_geo** is the orientation of the X_1 direction of the coordinate system that the measurement directions are in. It is assumed that X_2 is horizontal and X_3 follows the "right-hand rule " discussed in Chapter 3.

1) Use the program **mk_nfo** to enter the orientation data for the samples in the previous example. These are:

Sample	NB Azimuth (Strike)	NB Plunge (Dip)	Stratigraphic height
tn07a	250	22	15.6
tn07c	275	51	15.6
tn07d	245	42	15.6

Put the orientation data calculated by **mk_nfo** into a file called **ex6_c.nfo** These samples were taken with the "cube" method, so notebook (NB) values for "azimuth" and "plunge" are actually the strikes and dips of a perpendicular face (see Figure 3.2). The sediments are consistently oriented

with a strike of 240 and a dip of 20 (dip to the right of strike), but stratigraphic position is of interest, so these data should also be entered. The
samples were collected from the East African Rift valley at 1° N, 35° East
in 1985 and according to the results of the **igrf** program, the magnetic
declination was negligible (less than 1°).

2) Use the program **mk_nfo** to make an **.nfo** file for samples drilled from
a lava flow (listed below). The orientation device measures the azimuth of
the direction of drilling and the angle that the drill direction made with
the vertical down direction. Also available is sun compass information for
the azimuth of the drill direction. These data came from the Island of La
Palma in the Canary Islands at 29° N and 18° W on December 31, 1994.
The reference field for these data (from **igrf**) has a declination of 351°.

3) Although the program has built-in options for converting notebook azimuth and plunge data to the lab arrow, every paleomagnetist seems to
develop his or her own system. For this program, the "user defined" option
is to enter the following orientation data assuming that the notebook values
refer to the "up drill direction" as opposed to the "down drill direction"
(the direction of the drill arrow on Figure 3.1d would point in the opposite
direction).

Sample	NB Azimuth	NB Plunge
lp01a	192	-2
lp01b	221	-6
lp01c	296	-36
lp01d	160	-2
lp01e	135	-58

Solution

mk_nfo is a large and complicated program, so let us start with the help
message generated by the **mk_nfo -h** command:

Usage mk_nfo [-HbBDstfk][strike dip][magdec][lat. long. delta T][cpsdu]
[az_add az_mult pl_add pl_mult][basename][keyboard input]

    makes information file(s) for conversion of data into
      geographic, tectonic and stratigraphic references

Options:

    -H stratigraphic position data
    -b [strike dip] of bedding for each sample
    -B [strike dip] of bedding for entire suite
    -D add [magdec] to all declination/strike info
    -s sun compass data using [lat. long. dT]
      lat/long of study area
      and dT is time difference from GMT

-t type of conversion from

    notebook azimuth (= NBaz)/ notebook plunge (= NBpl) to

    lab azimuth (= LABaz)/ lab plunge (= LABpl)

    [c]ube:

        NBaz/NBpl are strike and dip on face perpendicular

           to face with lab arrow

        LABaz=NBaz-90, LABpl=NBpl-90

    [p]omeroy orientation device:

        NBaz/NBpl are direction of drill and angle from vertical

        LABaz=NBaz, LABpl=-NBpl

    [s]trike/dip:

        NBaz/NBpl are strike and dip on face with lab arrow

        LABaz=NBaz+90, LABpl=NBpl

    [d]rill direction/dip:

        NBaz/NBpl are direction of drill (az/pl) in plane perpendicular

           to face with lab arrow

        LABaz=NBaz, LABpl=NBpl-90

    [u]ser defined conversion

        input [az_add az_mult pl_add pl_mult] to

        specify affine for NBaz/NBpl to LABaz/LABpl

        e.g. for [p] above, az_add=0, az_mult=1

           pl_add=-90, az_mult=-1

-f output file specified as [basename]

    ouput file will be appended to basename.nfo and if

    sun compass data calculated, basename.snfo too

-k input from keyboard with prompts

    < cntl-D. to quit.

Input:

    sample,NBaz,NBpl [pos][strike dip][yyyymmdd hhmm shadow]

Output:

    sample, pos, type, NBaz,NBpl,LABaz,LABpl,NBstr,strike,dip

Defaults:

    read/write from Standard I/O

    input only: sample, NBaz,NBpl

no declination adjustment

1) In order to carry out the first part of the program, we can enter the data into a file with the fields:

sample NBaz NBpl height

and put in the orientation type (c), bedding orientation (240 20) as command line arguments:

% mk_nfo -HtBf c 240 20 ex6_c < ex6.mk_nfo.c

This creates a file named ex6_c.nfo that looks like this:

```
tn07a   15.6 c   250.   22.   160.   -68.   240.   240.   20.
tn07c   15.6 c   275.   51.   185.   -39.   240.   240.   20.
tn07d   15.6 c   245.   42.   155.   -48.   240.   240.   20.
```

Look at the "output" format in the help message for the key to what these fields mean.

2) To make a **.nfo** file for the La Palma samples, we need to specify the notebook Plunge: [0.0] that the type is '**p**', that the magnetic declination is -9° and that there are sun compass data from the correct latitude and longitude (29, -18) and the difference between the measurement time and Greenwich mean time, which in this case is 0. We can either put the data in a datafile, or enter the data in an interactive way (keyboard input). For the purposes of illustration, we will do it the interactive way. Data in brackets will be retained by pressing return.

```
% mk_nfo -ktDs p -9 29 -18 0
Sample name: [AB123a ]
   Control-D to quit
lp01a
   Notebook Azimuth: [ 0.0]
12
   Notebook Plunge: [ 0.0]
88
   yyyymmdd: [ 19560126]
19941231
   hhmm: [ 0300]
1314
   Shadow angle: [ 0.0]
359
   lp01a 0.0 c 12. 88. 3. -88. 0. -9. 0. [magnetic declination]
   lp01a 0.0 c 12. 88. 359. -88. 0. -9. 0. [sun compass dec.]
   Sample name: [lp01a ]
      Control-D to quit
lp01b
   Notebook Azimuth: [ 12]
41
   Notebook Plunge: [ 88]
84
   yyyymmdd: [ 19941231]
just hit return to keep this value!
   hhmm: [ 0300]
1318
   Shadow angle: [ 0.0]
28
```

lp01b 0.0 c 41. 84. 32. -84. 0. -9. 0.
lp01b 0.0 c 41. 84. 29. -84. 0. -9. 0.
    Sample name: [lp01b]
      Control-D to quit
and so on.
If we had specified an output file basename (e.g. **lp**), the program would
have created two files, **lp.nfo** and **lp.snfo**. The magnetic compass information goes into the **.nfo** file, while the sun compass data go into the **.snfo**
file. These are not always the same as you can see by comparing the LABaz
fields in the two examples shown above. The user must choose between the
two. One way to do this is to concatenate the two files using **cat** and then
sort them:
% **cat lp.nfo lp.snfo | sort > lp_all.nfo**
Then you must go through line by line and choose which method gives the
most reliable azimuth.
3) Finally, we must do the third part of the problem which is to supply
user defined affines for the NBaz and NBpl variables. With a little thought,
we find that, if NBpl is negative (we were drilling downwards), then we
must add 180 to the NBaz to get the LAB_az correctly, so az_add = 180
and az_mult = 1. The LABpl = -90 - NBpl = -1 (NBpl + 90), so pl_add =
90 and pl_mult = -1. If the drill direction was up, it is another whole ball
game. In this excercise, all the drill directions were down (NBpl < 0), so
we will use az_add = 180, pl_add = 90 and both multipliers = 1. Having
put the orientation data into a file **ex6.mk_nfo.u**, we type:
%**mk_nfo -t u 180 1 90 -1 < ex6.mk_nfo.u**
and get:

lp01a	0.0	u	192.	-2.	12.	-88.	0.	0.	0.
lp01b	0.0	u	221.	-6.	41.	-84.	0.	0.	0.
lp01c	0.0	u	296.	-36.	116.	-54.	0.	0.	0.
lp01d	0.0	u	160.	-2.	340.	-88.	0.	0.	0.
lp01e	0.0	u	135.	-58.	315.	-32.	0.	0.	0.

• **Example 6.4**
Use the program **mag_dat** to convert the file created in Example 6.2
(**ex6.vmag**) into what we will call a **.dat** format. Use the **.nfo** file created in Example 6.3 **ex6_c.nfo** to rotate the vectors into geographic and
tilt adjusted coordinates.
The syntax of **mag_dat** is simple; it reads from standard input, writes to
standard output and the **.nfo** format file is specified as a command line
argument, following the switch **-n**:
% **mag_dat -n ex6_c.nfo < ex6.vmag**
to which the response is:

tn07a	15.6	0.00	2.3	0.8420E-04	54.7	-0.7	55.3	-2.4
tn07a	15.6	150.00	2.4	0.6500E-04	46.4	-2.6	48.1	-7.1
tn07a	15.6	300.00	2.3	0.2930E-04	31.4	-7.5	35.3	-16.6
tn07a	15.6	400.00	1.7	0.1260E-04	19.5	-8.1	23.6	-20.6
tn07a	15.6	450.00	0.7	0.4140E-05	344.1	4.7	344.5	-14.7
tn07a	15.6	500.00	1.2	0.4190E-05	73.5	1.2	72.3	5.7
tn07a	15.6	550.00	1.1	0.3650E-05	353.5	-14.6	357.3	-32.7
tn07a	15.6	600.00	0.8	0.2230E-05	350.7	11.2	350.5	-7.6

.

.

.

The help message will inform you that these fields are:
sample, position, treatment, CSD, intensity, geographic D and I, tilt adjusted D, and I.

• Example 6.5

Suppose you have collected the AMS of a set of samples from two margins of a dike (nominally the east and west margins) shown in Table 6.1. These samples were measured and you calculated the $\bar{s}$ data using the techniques discussed in Chapter 5 and are listed in the table below. Several dike orientation measurements were made and the dip directions and dips were: (102/70; 98/67; 108/82). Also, some outcrop flow lineations were noted as declination inclination pairs: (20/38; 20/36; 24/30). Finally, there are two tectonic rotations necessary to restore the dike to its original position (strike # 1 = 270, dip # 1 = 10; strike # 2 = 205, dip # 2 = 20).
Use the program **plotdike** to analyze these data. Plot the bootstrapped eigenvectors instead of ellipses and use a site parametric bootstrap.
Solution
Place the eastern and western margin data in files called **e.s** and **w.s**, respectively. Put the dike orientations (dip direction, dip) in a file called **dike.dd** and the lineation data (azimuth, plunge) in **lin.di**. Put the structural corrections (strike, dip, strike, dip) in a file called **struct.dat**. It is best if each dike has its own directory (in this example, it is called **dike_example**) in order to avoid confusion and overwriting of files.
Now type:
% **plotdike -Pv | plotxy**
The resulting **mypost** file is shown in Figure 6.28.

TABLE 6.1. Data for Example 6.5

		Data for eastern margin			
s_1	s_2	s_3	s_4	s_5	s_6
0.33355311	0.33144355	0.33500329	0.00173934	0.00264716	0.00239888
0.33409184	0.33231828	0.33358982	0.00194153	0.00279057	0.00198475
0.33422312	0.33073765	0.33503917	0.00210855	0.00211220	0.00293344
0.33278510	0.33232915	0.33488578	0.00160304	0.00229841	0.00158643
0.33341962	0.33295918	0.33362123	0.00072081	0.00142357	0.00170842
		Data for western margin			
s_1	s_2	s_3	s_4	s_5	s_6
0.33382347	0.33262473	0.33355179	0.00067366	0.00040422	0.00137196
0.33473068	0.33085340	0.33441597	0.00083552	0.00052653	0.00116593
0.33418971	0.33097920	0.33483109	0.00020041	0.00042602	0.00087438
0.33415532	0.33155510	0.33428958	0.00043263	-.00015507	0.00020517
0.33509848	0.33025289	0.33464867	0.00054253	0.00053096	0.00126909

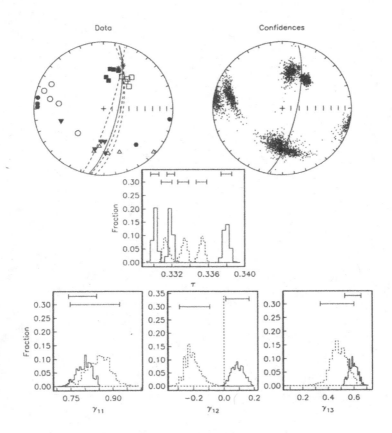

Figure 6.28.   Output for Example 6.5, generated by the command **plotdike -Pv |**
plotxy from within the directory **dike_example**. Conventions are the same as in Figure
6.21.

APPENDIX 1

THE PMAG SOFTWARE PACKAGE

1.1. Getting started

After each chapter in the book, there are a number of practical examples
for paleomagnetic problems using real data. The programs referred to in
the examples are part of the PMAG package of programs. The distribution
described here is **pmag1.7**. The most current version of the PMAG package
is available by anonymous ftp at:

ftp://sorcerer.ucsd.edu/pub/pmag.

These programs were designed to operate in a UNIX type environment
and we include the source code (in Fortran 77) as well as compiled ver-
sions for Linux, Solaris, and Mac OS X operating systems. There are also
executables (available for pmag1.4 only) provided for running under the
MSDOS prompt on a PC. There are many advantages to working in a
UNIX environment so we will start with a brief discussion of how to get
along in UNIX in general. There are many excellent books on UNIX and
the reader is referred to them for a detailed discussion. What follows are
the barest essentials for being able to use the programs in this book.

1.1.1. SURVIVAL UNIX

This book assumes that you have an account on a UNIX type machine and
know how to log in. If you are using a Macintosh with Mac OS X, just use
the terminal window. In this book, what *you* type is printed in boldface.
At the end of every line, you must also type a carriage return ("Return" or
"Enter" on most computer keyboards). To end input into a program, press
the control key and "d" $< control - D >$ simultaneously.

After a login message specific to your computer, you will get a command
line prompt which varies widely. In this book, we use the symbol "%" as
the command line prompt.

1.1.2. THE UNIX FILE STRUCTURE

Fundamental to the UNIX operating system is the concept of directories
and files. On windows-based operating systems, directories are depicted as
"folders" and moving about is accomplished by clicking on the different
icons. In UNIX, the directories have names and are arranged in a hierar-
chical sequence with the top directory being the "root" directory, known as
"/" (see Figure A.1. Within the "/" directory, there are subdirectories (e.g.
usr and **home**). In any directory, there can also be "files" (e.g. *ex1.1, ex1.2*
in the figure). Files can be "readable", "writable" and/or "executable".

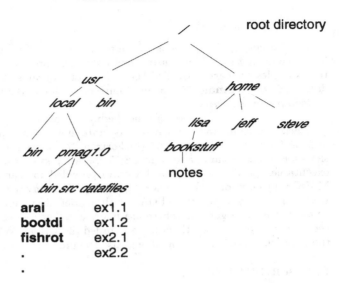

Figure A.1. Part of a UNIX filesystem tree. Directories are shown in italics, ascii files in plain text and executable programs in boldface.

When you log in you enter the computer in your "home" directory. To refer to directories, UNIX relies on what is called a "pathname". Every object has an "absolute" pathname which is valid from anywhere on the computer. The absolute pathname always begins from the root directory /. So the absolute pathname to the home directory **lisa** in Figure A.1 is **/home/lisa**. Similarly, the absolute pathname to the directory containing pmag1.7 executables is **/usr/local/pmag1.7/bin**. There is also a "relative" pathname, which is in reference to the current directory. If user "lisa" is sitting in her home directory, the relative pathname for the file **notes** in the directory **bookstuff** would be **bookstuff/notes**. When using relative pathnames, it is useful to remember that ./ refers to the current directory and ../. refers to the directory "above".

Commands typed at the command line prompt are handled by a program called the "shell". There are many different sorts of shells (e.g. *sh, csh, jsh, ksh, bash, tsh*) that have a different look and feel but they all perform the task of interpreting between the user and the "kernel" which is the actual UNIX operating system. In order to execute a command, the shell needs to know where the command is. There are several "built-in" commands, but most are programs that are either part of the operating system, or something someone wrote (like the ones referred to here). There are any number of places where programs are kept, so the shell looks in particular places determined by your "path" environment variable. To instruct the shell to look in directories other than the default directories (for example in **/usr/local/pmag1.7/bin**), ask your system administrator to add this directory to your "path". Otherwise, you can always type the absolute pathname for any program (e.g. **/usr/local/pmag1.7/bin/fishrot** to execute the program **fishrot**.

1.1.3. REDIRECTING INPUT AND OUTPUT

Most UNIX programs print output to the screen and read input from the keyboard. This is known as "standard input and output" or "standard I/O" in the following. One of the nicest things about UNIX is the ability to redirect input and output. For example, instead of typing input to a program with the keyboard, it can be read from a file using the symbol <. Output can either be put into a file using the symbol >, appended to the end of a file with >> or used as input to another program with the UNIX pipe facility (|).

1.1.4. WILDCARDS

UNIX has the ability to refer to a number of files and/or directories using "wildcards". The wildcard for a single character is "?" and for any number of characters is "*". For example, to refer to all the files beginning with "ex" in the directory:
/usr/local/pmag1.7/datafiles,
we use:
/usr/local/pmag1.7/datafiles/ex*.
To refer only to those from Chapter 1, we use:
/usr/local/pmag1.7/datafiles/ex1.* .

1.1.5. UNIX COMMANDS

Now we briefly describe essential UNIX commands.
• **awk**
Usage: awk [options] [file(s)] [Standard I/O]
Description: There are whole books on this program. We will use **awk** in a very primitive way to select specific columns from standard input for use as input into another program. For example, if the third and fourth column of file **myfile** is desired as input to program **myprog** type:
% awk '{print $3, $4 }' myfile | myprog
• **cat**
Usage: cat [options][file(s)] [Standard I/O]

Description: Concatinates and displays files. It reads from standard input or from the specified file(s) and displays them to standard output.

• **cd**

Usage: cd [directory]

Description: Changes directory from current directory to the one specified.

• **cp**   •

Usage: cp [file1] [file2]

Description: Copies files or directories.

• **grep**

Usage: grep [options] [expression] [file(s)]

Description: **grep**, like **awk**, is a very versatile (and complicated) program about which one could read an entire book. However, we will use **grep** simply to pick out particular key strings from a file. For example, if a file **myfile** contains lines of data for many samples, and we wish to consider the data for a single sample **mysamp**, lines containing the word **mysamp** can be "grepped" out by:

% **grep "mysamp" myfile**

and listed to the screen, redirected to a file, or piped to a program.

• **join**

Usage: join [options] file1 file2

Description: *file1* and *file2* share a common "join" field, by default the first column. This could for example be a sample name. The output file prints the join field, followed by the rest of the line from **file1**, then the rest of the line in **file2**. Say **file1** has magnetometer output data, with lines of data containing: sample, treatment, D, I, M and **file2** has pertinent information such as sample orientation, with lines: sample, azimuth, plunge. We may wish to attach the sample orientation data to the magnetometer output for further processing. **join** allows us to do this by the following:

% **join file1 file2**

• **ls**

Usage: ls [options] [directory name]

Description: Lists the contents of the specified directory. If none specified, lists the current directory.

• **man**

Usage: man [command name]

Description: Prints the on-line documentation for the specified command.

• **mkdir**

Usage: mkdir [directory name]

Description: Makes a directory with the specified name.

• **more**

Usage: more [file name]

Description: Displays the contents of a text file on the terminal, one screenful at a time. To view line by line, type RETURN. To view page by page, use the space bar.

• **mv**

Usage: mv [file1 file2]

Description: Renames *file1* to *file2*. This also works for directory names.

• **paste**

Usage: paste [options] [file1 file2]

Description: "Pastes" lines from *file2* onto the corresponding line in *file1*.

• **pwd**

Usage: pwd

Description: Prints the absolute pathname of the working (or current) directory.

• **rm**

Usage: rm [options] [file(s)]

Description: Deletes the specified file or files.

• **rmdir**

Usage: rmdir [options] [file(s)]

Description: Deletes the specified directories. Note: the directory has to be empty.

• **tee**

Usage: tee [file name]

Description: Makes a copy of the standard input to the specified file, then passes it to standard output.

More detailed descriptions are usually available on-line with the **man** command. For example, to find out more about **cat**, type:

% **man cat**

and read all about it.

1.1.6. TEXT EDITORS

Text editing is a blessing and a curse in most UNIX systems. You either love it or hate it and in the beginning, you will certainly hate it. There are many ways of editing text and the subject is beyond the scope of this book. Almost all UNIX systems have some flavor of **vi** so try reading the man pages for that.

1.2. Getting the programs used in this book

The PMAG software distribution contains the fortran 77 source code, compiled programs for the Solaris, Linux and Mac OS X operating systems as well as executables for MSDOS (only the pmag1.4 distribution for MSDOS).

To run the compiled programs, simply "untar" them using for example the
command

%**tar xvf macosx.tar**

Put the resulting *macosx* directory in your path or move the files in that
directory to a directory that is in your path already.

We recommend that you "untar" the file *pmag1.7.tar*. This contains the
source code and miscellaneous files for use with the the exercises in this
book. Follow these instructions:

1) Place **pmag1.7.tar** in any directory (PARENTDIR) that you have write
permission for.

Your fortran compiler must be known as "f77" or change the make.all
file for the name of your compiler (eg., fort77).

2) Execute the following commands:

%**tar xf pmag1.7.tar**

%**cd pmag1.7/src**

3) Type:

%**source make.all**

Make sure **PARENTDIR/pmag1.7/bin** is in your path (or copy them
to a directory that is in your path) and try out the programs as described
in the book. The data files mentioned in the excercises can be found in
PARENTDIR/pmag1.7/datafiles with names is given in the book.

1.3. Other necessary software

Included in this package of programs is a program called **plotxy**, written
by Robert L. Parker and Loren Shure. This is freeware and is supplied as
is. For a detailed description on how to use it, print out the file:

pmag1.7/src/Plotxy/plotxy.doc.

The program **plotxy** reads commands from standard input and creates
a postscript file named by default *mypost*. Finally, you will need to have
some sort of Postcript viewer, e.g. **pageview, ghostscript, ghostview**
or commercial illustration product OR just convert the postscript files to
pdf format and view them with Acrobat Reader available for free at this
website:

http://www.adobe.com

1.4. Description of programs in excercises

The software described here is updated occasionally. For the description of
the program you have, type:

% **program_name -h**

for a help message.

Here follows are hints for the use of all the programs referred to in the excercises in this book as well as some supplemental programs that are helpful.

The programs described here are part of the package **pmag1.7**. They are designed to take advantage of the power of UNIX in that they all take command line arguments. The common features are:

1) All programs respond with a help message if the [-**h**] switch is used:

% **program_name -h**

2) Programs that produce pictures use the freeware program **plotxy** (see section on getting and installing **plotxy**. PMAG programs output **plotxy** commands which can be piped directly to **plotxy**:

% **program_name | plotxy**

This causes **plotxy** to create a postscript file, by default *mypost* which can be viewed with a postscript viewer such as **pageview**, **ghostscript** or **ghostview**, or it can be printed to a postscript compatible printer. It can also be converted to *pdf* format. The **plotxy** commands generated by the PMAG programs can also be saved to a file and modified as needed:

% **program_name > myfile.com**

A text editor can be used to modify the commands to taste (see **plotxy** documentation **plotxy.doc** in the Plotxy directory:

/usr/local/pmag1.7/Plotxy),

or online at:

http://sorcerer.ucsd.edu/pmag/plotxy.html.

3) Programs read from standard input and write to standard output unless otherwise noted. All input files are expected to be space delimited - not tab or comma delimited. The typical syntax will be:

% **program_name < input_file > output_file**

The output from one program can be piped as input to another program. Here follows a brief manual for using the programs in the PMAG package. Refer to the examples at the end of each Chapter for further hints about how to use them.

● **arai**

Usage: arai [-sfmd][min max field][Standard I/O]

Description: makes an Arai plot from input data.

Example 3.5

Options:

   -s sets fit from [min] to [max]

   -f sets lab field to [field] (in Tesla)

   -m uses .mag file as input

Defaults:

   finds "optimum" interval - beware! You may have to "tweak" the [min] and [max] values using the [-s] switch

default laboratory field is 40 μT.

Input options:

1em Default input:

  Sample treatment intensity D I

.mag file [-m] option:

  Sample treatment CSD intensity D I

  treatment steps are coded as follows:

    XXX.YY where XXX is the temperature and

    YY is as follows:

    NRM data: .00

    pTRM: .11

    pTRM check: .12

Output: plotxy commands.

- **bootams**

Usage: bootams [-pP] [Standard I/O]

Description: calculates bootstrap eigenparameters from input file.

Example: 5.9

Options:

  -p option specifies parametric (by sample) bootstrap

  -P option specifies parametric (by site) bootstrap

Input: s_1 s_2 s_3 s_4 s_5 s_6 σ

Output: bootstrap error statistics:

  $\tau_1 \; \sigma \; D \; I \; \eta \; D_\eta \; I_\eta \; \zeta \; D_\zeta \; I_\zeta$

  $\tau_2 \; \sigma \; D \; I \; \eta \; D_\eta \; I_\eta \; \zeta \; D_\zeta \; I_\zeta$

  $\tau_3 \; \sigma \; D \; I \; \eta \; D_\eta \; I_\eta \; \zeta \; D_\zeta \; I_\zeta$

- **bootdi**

Usage: bootdi [-pPv] [Standard I/O]

Description: calculates bootstrap statistics for input file.

Example: 4.9

Options:

  -p option selects parametric bootstrap

  -P works on principal eigenvectors

  -v spits out bootstrapped means

Defaults:

  simple bootstrap

  works on Fisher means

Input: D I $[\kappa \; N]$

Output:

  Fisher statistics if Fisherian, otherwise bootstrap ellipses for one or two modes:

    Mode $\eta, D_\eta I_\eta \zeta D_\zeta I_\zeta$.

  or if -v selected, bootstrapped eigenvectors DI

- **bootstrap**

Usage: bootstrap [-pb] [Nb] [Standard I/O]

Description: Calculates bootstrap statistics from input data.

Example: 4.8

Options:

  -p plot a histogram

  -b sets number of bootstraps [Nb](<10000)

Input: single column of numbers

Output:

  if no plot, then:

  N Nb mean bounds_containing_95%_of_means

  if plot, then output is series of plotxy commands

Defaults:

  no plot and nb=1000

- **cart_dir**

Usage: cart_dir [Standard I/O]

Description: Converts cartesian data to geomagnetic elements

Example: 1.2

Input: x_1 x_2 x_3

Output: D I magnitude

- **cart_hist**

Usage: cart_hist [-dcbpr][dec inc][file1 file2][Standard I/O]

Description: Makes histograms of cartesian coordinates of input.

Example: 4.11

Options:

  -d compares with direction [dec inc]

  -c compares two files [file1 file2]

  -b plots confidence bounds

  -p specifies parametric bootstrap

  -r flips second mode for reversals test

Input: D I $[\kappa$ $N]$

Output is plotxy command file

Defaults:

  standard input of single file

  no confidence bounds

  simple bootstrap

  no reversals test

- **curie**

Usage: curie -[lspt] [smooth] [low hi step] [Tmin Tmax] [Standard I/O]

Description: Analyzes Curie temperature data.

Example: 2.2

Options:

-l smooth over [smooth] data points
  NB: [smooth] must be an odd number ≥ 3
-s scan range of smoothing intervals
  [low] to [hi] using a spacing of [step]
  [low],[hi] and [step] must be odd
-p plot option on to generate Plotxy command file
    can be piped directly to plotxy and viewed:
  curie -p < filename | plotxy; ghostview mypost
    printed:
  curie -p < filename | plotxy; lpr mypost
    or saved to a file for modification:
  curie -p < filename > eqarea.com
-t truncates to interval between [Tmin] and [Tmax]
input:
  temperature,magnetization
Defaults:
  no smoothing
  plot option off
  uses entire record
- **di_geo**
Usage: di_geo [Standard I/O]
Description: Rotates directions from specimen to geographic coordinates.
Example: 3.2
Input: D I azimuth plunge
Output: D I (in geographic coordinates)
Notes: the azimuth and plunge are the declination and inclination of the arrow used for reference during the measurements.
- **dir_cart**
Usage: dir_cart [-m] [Standard I/O]
Description: Converts geomagnetic elements to cartesian coordinates.
Example: 1.1
Option: -m read magnitude field
Input: D I [magnitude]
Output: x_1 x_2 x_3
Notes: VGP longitude, latitude can be substituted for D, I.
- **di_tilt**
Usage: di_tilt [Standard I/O]
Description: rotates directions from geographic to tilt adjusted coordinates.
Example: 3.2
Input: D I strike dip
Output: D I (in adjusted coordinates)
Notes: convention is that dip is to the "right" of strike.

- **di_vgp**

Usage: di_vgp [Standard I/O]

Description: Transforms declination/inclination to VGP.

Example: 1.5

Input: D I lat.($°$ N) long.($°$ E)

Output: Pole_Longitude Pole_Latitude

Notes: convention is positive: North, negative: South and positive: East, negative: West

- **eigs_s**

Usage: eigs_s [Standard I/O]

Description: Converts eigenparameters to .s format.

Example: 5.2

Input:

$$\tau_3\ D_{\mathbf{V}_3}\ I_{\mathbf{V}_3}\ \tau_2\ D_{\mathbf{V}_2}\ I_{\mathbf{V}_2}\ \tau_1\ D_{\mathbf{V}_1}\ I_{\mathbf{V}_1}$$

Output:s_1 s_2 s_3 s_4 s_5 s_6

Notes: τ_1 is the largest eigenvalue and $\mathbf{V}_i$ are the associated eigenvectors.

- **eqarea**

Usage: eqarea [Standard I/O]

Description: Makes an equal area projection of input data.

Example: 1.3

Input: D I

Output: plotxy commands

- **fishdmag**

Usage: fishdmag [-fdm] [beg end ta] [Standard I/O]

Description: Calculates Fisher mean from specified portion of demagnetization data.

Example: 4.3

Options:

  -f Fisher mean from [beg] to [end] steps

  -d uses .dat file as input

  if [ta] = 0 (default), uses geographic (D_g, I_g)

  if [ta] = 1 uses tilt adjusted (D_{ta}, I_{ta})

  -m uses .mag file as input

where [beg] and [end] are the number of the treatment (i.e. 1st, 2nd, 3rd).

Input options:

Default input:

  Sample treatment intensity D I

.mag file [-m] option:

  Sample treatment CSD intensity D I

.dat file [-d] option

  Sample position treatment CSD intensity D_g I_g D_{ta} I_{ta}

Output: Sample f n beg end α_{95} $\bar{D}$ $\bar{I}$

- **fisher**

Usage: fisher -kns [kappa] [N] [seed] [Standard I/O]

Description: generates set of Fisher distribed data from specified distribution.

Example: 4.1

Options:

-k specifies κ as [kappa]

-n specifies number as [N]

-s specifies random seed (non-zero integer) as [seed]

Defaults:

$\kappa = 30$

$N = 100$

seed $= 1200$

- **fishqq**

Usage: fishqq [Standard I/O]

Description: plots Q-Q diagram for input against Fisher distribution.

Example: 4.5

Input: DI

Output: plotxy commands

- **fishrot**

Usage: fishrot -kndis [kappa] [N][dec][inc][seed] [Standard I/O]

Description: draws a Fisher distribution with mean of [dec] [inc] and [kappa], [N], using random seed [seed].

Example: 4.4

Options:

-k specifies κ as [kappa]

-n specifies number as [N]

-d specifies D as [dec]

-i specifies I as [inc]

-s specifies [seed] for random number generator (non-zero)

Defaults:

$\kappa = 30$

$N = 100$

$D = 0$

$I = 90$

- **foldtest**

Usage: foldtest [-p] [Standard I/O]

Description: Performs bootstrap fold test.

Example: 4.13

Options:

-p option selects parametric bootstrap

Input: $D\ I$ strike dip $[\kappa\ N]$

Output: plotxy commands

Notes: unfolding curve of data is solid line, pseudo samples are dashed histogram is fraction of τ_1 maxima. Also, dip is to the "right" of strike.

- **gauss**

Usage: gauss -msni [mean] [sigma] [N] [seed] [Standard I/O]

Description: draws a set of Gaussian distributed data from specified distribution.

Example: 4.6

Options:

  -m sets the mean to [mean]

  -s sets the standard deviation to [sigma]

  -n sets the number of points to [N]

  -i sets integer random seed to [seed]

Defaults:

  [mean] is 1

  [sigma] is .5

  [N] is 100

- **gofish**

Usage: gofish [Standard I/O]

Description: calculates Fisher statistics from input file.

Example: 4.2

Input: $D\ I$

Output: $\bar{D}\ \bar{I}\ N\ R\ k\ \alpha_{95}$

- **goprinc**

Usage: goprinc [Standard I/O]

Description: calculates principal component from input data.

Example: 4.2

Input: $D\ I$

Output: $D\ I\ N\ \tau_1$

- **gtcirc**

Usage gtcirc [-gdm] [beg end ta] [Standard I/O]

Description: calculates best-fit plane through specified input data.

Example: 3.3

Options:

-g best-fit great circle (plane) from [beg] to [end] steps

  [beg] and [end] are the numbers of the treatment step.

  For example the kNRM step is [1], the second step is [2], etc.

-d uses .dat file as input

  if [ta]=0 (default), uses geographic (D_g, I_g)

  if [ta] = 1 uses tilt adjusted (D_{ta}, I_{ta})

-m uses .mag file as input

Input options:

Default input:

    Sample treatment intensity D I

.mag file [-m] option:

    Sample treatment CSD intensity D I

.dat file [-d] option

    Sample position treatment CSD intensity D_g I_g D_{ta} I_{ta} Output:

Sample g N beg end MAD D I

where D and I are for the pole to the best-fit plane

and [beg] and [end] are the numbers of the treatment step.

- **histplot**

Usage: histplot [-lb] [bin] [Standard I/O]

Description: creates a histogram of input data.

Example: 4.6

Options:

-l plots the distributions of logs

-b sets bin size to [bin]

Input: single column of data

Output: Plotxy commands

Defaults:

    not logs

    auto binning

- **hystcrunch**

Usage: hystcrunch -[mpatl] [trunc_deg] [label] [Standard I/O]

Description: plots hysteresis loop data and massages them.

Example: 2.3

Options:

    -m Micromag data file

    -p do not plot

    -a do not adjust slope

    -t truncate to trunc_deg harmonics

    -l label plot with label

Defaults:

    - xy data file

    - retain 99 terms of FFT

    - adjust for high field slope

    - no plot label

    - generate plotxy commands

- **igrf**

Usage: igrf [Standard I/O]

Description: calculates reference field vector at specified location and time
uses appropriate IGRF or DGRF for date > 1945.

Example: 1.4

Input: year altitude latitude longitude
where year is decimal year, altitude is in kilometers, latitude is in ° N and
longitude is in ° E.
Output: D I B (nT)

• **incfish**
Usage: incfish [Standard I/O]
Description: calculates an estimated inclination, assuming a Fisher distri-
bution, for data with only inclinations. Uses the method of McFadden and
Reid [1982].
Example: 4.4
Input: inclinations
Output: $< I >$, upper and lower bounds, N, estimated κ and α_{95}.

• **jackstrat**
Usage: jackstrat Standard I/O
Description: calculates magnetostratigraphic jackknife parameter J.
Example: 6.1
Input: VGP latitudes or inclinations in stratigraphic order
Output: plotxy commands

• **k15_hext**
Usage: k15_hext [-tga] [Standard I/O]
Description: calculates Hext statistics from 15 measurements uses Jelinek's
15 measurement scheme.
Examples: 5.5 & 5.6 .
Options:
   -a average whole file
   -g geographic coordinates
   -t geographic tilt coordinates
Default: average by sample
Input: 1 line with sample name, [azimuth, plunge, strike, dip] followed by
3 rows of 5 measurements for each specimen in the following order (see
Chapter 5):
   K_1 K_2 K_3 K_4 K_5
   K_6 K_7 K_8 K_9 K_{10}
   K_{11} K_{12} K_{13} K_{14} K_{15}
Output: Hext statistics
   [if individual samples, id and bulk chi]
   F F_{12} F_{23}
   τ_1 D I ϵ_{12} D I ϵ_{13} D I
   τ_2 D I ϵ_{12} D I ϵ_{13} D I
   τ_3 D I ϵ_{12} D I ϵ_{13} D I

• **k15_s**
Usage: k15_s [-gt] [Standard I/O]

Description: calculates $\bar{s}$ from 15 measurements scheme (K_i) (see Chapter 5).

Example: 5.4

Options:

  -g geographic rotation

  -t geographic AND tectonic rotation

Input: (see k15_hext)

Output: $\bar{s}_1$ $\bar{s}_2$ $\bar{s}_3$ $\bar{s}_4$ $\bar{s}_5$ $\bar{s}_6$ σ

• lnp

Usage: lnp [-f] [infile] [Standard I/O]

Description: calculates Fisher mean from combined directed lines and great circles using the method of McFadden and McElhinny [1988].

NB: this program has changed somewhat since the first distribution!

Example: 4.3

Options:

  -f calculates mean from data in [infile], one site at a time

Input: output file from **pca**, **gtcirc**, **fishdmag** programs, i.e.

  Sample [fpg] n beg end α_{95} MAD D I

  sample name convention: ABC123D[1]

  where ABC is a study designator of letters (any length)

  123 is the site number (any length)

  D is a (single) letter for each separately oriented sample

  [1] is an optional (single digit) specimen number

Default: assumes whole file is one site.

Output:

  site NL $NG\kappa\alpha_{95}$ D I

  where NL is the number of directed lines

  where NG is the number of great circles

• mag_dat

Usage: mag_dat -n nfofile [Standard I/O]

Description: Converts magnetometer data file to format with geographic and tilt adjusted coordinates. Also pastes in stratigraphic position data. Uses **.nfo** files made by **mk_nfo**.

Example: 6.4

Input:

  magnetometer data (.mag file format):

  sample[optional specimen number] treatment CSD intensity D I

  (.nfofile format):

  Sample position [cpsdu] NBaz NBpl LABaz LABpl NBstr strike dip

Output:

  (.dat file foramt)

    Sample position treatment CSD intensity D_g I_g D_{ta} I_{ta}

where D_g and I_g are D, I in geographic coordinates and D_{ta} and I_{ta} are D, I in tilt adjusted coordinates.

• **mk_nfo**

Usage: mk_nfo [-HbBDstfk][strike dip][magdec][lat. long. delta T][cpsdu][az_add az_mult pl_add pl_mult][basename][keyboard input]

Description: Makes an information file(s) for conversion of data into geographic, tectonic and stratigraphic references.

Example: 6.3

Options:

-H stratigraphic position data

-b structural strike/dip for each sample

-B structural for entire suite read as [strike dip]

-D add [magdec] to all declination/strike info

-s sun compass data using [lat. long. dT]
   lat/long of study area
   and dT is time difference from GMT

-t type of conversion from
   notebook azimuth (= NBaz)/ notebook plunge (= NBpl) to
   lab azimuth (= LABaz)/ lab plunge (= LABpl)
   [c]ube:
      NBaz/NBpl are strike and dip on face perpendicular
      to face with lab arrow
      LABaz=NBaz-90, LABpl=NBpl-90
   [p]omeroy orientation device:
      NBaz/NBpl are direction of drill and angle from vertical
      LABaz=NBaz, LABpl=-NBpl
   [s]trike/dip:
      NBaz/NBpl are strike and dip on face with lab arrow
      LABaz=NBaz+90, LABpl=NBpl
   [d]rill direction/dip:
      NBaz/NBpl are direction of drill (az/pl) in plane perpendicular
      to face with lab arrow
      LABaz=NBaz, LABpl=NBpl-90
   [u]ser defined conversion
      input [az_add az_mult pl_add pl_mult] to
      specify affine for NBaz/NBpl to LABaz/LABpl
      e.g. for [p] above, az_add=0, az_mult=1
   pl_add=-90, az_mult=-1

-f output file specified as [basename]
   ouput file will be appended to basename.nfo and if
   sun compass data calculated, basename.snfo

-k input from keyboard with prompts

<cntl-D>. to quit.

Input:

Sample NBaz NBpl [position][strike dip][yyyymmdd hhmm shadow]

Output:

Sample position type NBaz NBpl LABaz LABpl NBstr strike dip

Defaults:

read/write from Standard I/O

input only: sample, NBaz,NBpl

no declination adjustment

LABaz=NBaz; LABpl=NBpl

● **pca**

Usage: pca [-pmd] [beg end][ta] [Standard I/O]

Description: Calculates best-fit line through specified input data.

Example: 3.3

Options:

-p PCA from [beg] to [end] steps

[beg] and [end] are the numbers of the treatment step.

For example the NRM step is [1], the second step is [2], etc.

-d uses .dat file as input

if [ta]= 0 (default), uses geographic (D_g, I_g)

if [ta] = 1 uses tilt adjusted (D_{ta}, I_{ta})

-m uses .mag file as input

Input options:

Default input:

Sample treatment intensity D I

.mag file [-m] option:

Sample treatment CSD intensity D I

.dat file [-d] option

Sample position treatment CSD intensity D_g I_g D_{ta} I_{ta}

Output:

Sample p N beg end MAD D I

where D and I are for the principal component

● **plotams**

Usage: plotams [-BpPvxjn] [name] [Standard I/O]

Description: Plots AMS data from s data

Example: 5.10

Options:

-B do not plot simple bootstrap ellipses

-p plot parametric (sample) bootstrap ellipses

-P plot parametric (site) bootstrap ellipses

-v plot bootstrap eigenvectors - not ellipses

-x plot Hext [1963] ellipses

   -j plot Jelinek [1978] ellipses
   -n use [name] as plot label
Default: plot only the simple bootstrap
Input: $s_1\ s_2\ s_3\ s_4\ s_5\ s_6\ [\ \sigma]$
Output: plotxy commands
• **plotdi**
Usage: plotdi [-pPv] [Standard I/O]
Description: Makes equal area plot of input data, with uncertainties.
Example: 4.10
Options:
   -p parametric bootstrap
   -P works on principal eigenvector
   -v plots bootstrapped eigenvectors
Defaults:
   simple bootstrap
   works on Fisher means
   plots estimated 95% conf. ellipses
Input: $D\ I\ [k\ N]$
Output: plotxy commands
• **plotdike**
Usage: plotdike [-BpPvn] [name] [Standard I/O]
Description: Makes a plot of ams data for dike margins. Designed for esti-
mating flow directions.
Example 6.5
   -B DONT plot simple bootstrap ellipses
   -p plot parametric (sample) ellipses
   -P plot parametric (site) ellipses
   -v plot bootstrap eigenvectors
   -n use [name] as plot label
Input:
   -one or both files called:
     e.s and w.s containing:
     $s1\ s_2\ s_3\ s_4\ s_5\ s_6\ [\sigma]$ for the nominal east and west margins
     respectively
   - a file called dike.dd containing one or more
     measurements of the dip direction and dip of the dike
   -optional files:
     struct.dat: contains first and second
        tectonic corrections as strike and dips
     lin.di: contains dec,inc of lineation data
Output: plotxy commands and a file:
   fort.20 is summary file

Default: plot only the simple bootstrap confidence ellipses
- **plotdmag**

Usage: plotdmag [-pgfrmd] [beg end][D][ta][Standard I/O]

Description: Makes orthogonal and equal area projections of input demagnetization data.

Example: 3.3

Options:

  -p PCA from [beg] to [end] steps

  -g gtcirc from [beg] to [end] steps

  -f fisher mean from [beg] to [end] steps

  -r plot horizontal axis = [D] degrees

  -m uses .mag file format

  -d uses .dat file format

Defaults:

  North on horizontal ([D] = 0)

  no PCA, gtcircle, or fisher calculations

Input options:

Default input:

  Sample treatment intensity D I

.mag file [-m] option:

  Sample treatment CSD intensity D I

.dat file [-d] option

  Sample position. treatment CSD intensity D_g I_g D_{ta} I_{ta}

Output: plotxy commands

Notes: select either p, f OR g

- **pseudot**

Usage: pseudot [-sm][min][ta][Standard I/O]

Description: Analyses pseudo-Thellier data

Example 3.6

Input options:

  -s sets the minimum field to [min]

  -m sets input file to .mag format

Input options:

Default input:

  Sample tr int dec inc

  .mag file option

  Sample tr csd int dec inc

treatment steps are coded as follows:

  XXX.XY where XXX.X is the AF level and

  Y is as follows:

  NRM data: 0

  ARM: 1

Output: plotxy command file
- **qqplot**

Usage: qqplot [Standard I/O]

Description: Plots data against Normal Quantile

Example: 4.7

Input: single column of data

Output: plotxy commands

Notes: On the plot, there are these parameters:

N: the number of data points

mean: the Gaussian mean

σ: the standard deviation

D: the Kolmogorov-Smirnov D-statistic

D_c: the critical value given N at 95% confidence

(if $D > D_c$, distribution is not Gaussian at 95% confidence)

- **s_eigs**

Usage: s_eigs [Standard I/O]

Description: Converts s format data to eigenparameters.

Example: 5.1

Input: s_1 s_2 s_3 s_4 s_5 s_6

Output:

$$\tau_1 \; D_{\mathbf{V}_1} \; I_{\mathbf{V}_1} \; \tau_2 \; D_{\mathbf{V}_2} \; I_{\mathbf{V}_2} \; \tau_3 \; D_{\mathbf{V}_3} \; I_{\mathbf{V}_3}$$

Notes: $D_{\mathbf{V}}$, $I_{\mathbf{V}}$ are the directions of the eigenvectors corresponding to the eigenvalues, τ and τ_1 is the largest eigenvalue.

- **s_flinn**

Usage: s_flinn [-pl] [Standard I/O]

Description: Plots Flinn diagram of input s data.

Example: 5.13

Options:

-p parametric bootstrap

-l plots L' versus F'

Input: s_1 s_2 s_3 s_4 s_5 s_6 $[\sigma]$

Output: plotxy commands

- **s_geo**

Usage: s_geo [Standard I/O]

Description: Rotates s data to geographic coordinates

Example: 5.3

Input: s_1 s_2 s_3 s_4 s_5 s_6 azimuth plunge

Output: s_1 s_2 s_3 s_4 s_5 s_6 rotated to geograpphic coordinates

- **s_hext**

Usage: s_hext [Standard I/O]

Description: Calculates Hext statistics from input s data.

Example: 5.7

Input: s_1 s_2 s_3 s_4 s_5 s_6
Output: Hext statistics

F F_{12} F_{23} σ

τ_1 D I ϵ_{12} D I ϵ_{13} D I
τ_2 D I ϵ_{21} D I ϵ_{23} D I
τ_3 D I ϵ_{32} D I ϵ_{31} D I

• s_hist

Usage: s_hist [-cbpP123t][file1 file2][standard I/O]
Description: Plots histograms of bootstrapped eigenparameters of s data.
Example: 5.11
Options:

-c compares [file1] and [file2]
-b plot 95% confidence bounds
-p sample parametric bootstrap
-P Site parametric bootstrap
-1 plots principal eigenvector
-2 plots major eigenvector
-3 plots minor eigenvector
-t plots eigenvalues

Defaults:

simple bootstrap
all eigenparameters
no confidence limits

Input: s_1 s_2 s_3 s_4 s_5 s_6 $[\sigma]$
Output: plotxy command for histograms of eigenvalues and eigenvectors of bootstrap samples

• s_jel78

Usage: s_jel78 [Standard I/O]
Description: Calculates Jelinek (1978) statistics from s data.
Example: 5.8
Input: s_1 s_2 s_3 s_4 s_5 s_6
Output: Jelinek statistics

τ_1 D I ϵ_{12} D I ϵ_{13} D I
τ_2 D I ϵ_{21} D I ϵ_{23} D I
τ_3 D I ϵ_{32} D I ϵ_{31} D I

• s_pt

Usage: s_pt [-p] [Standard I/O]
Description: Makes a Jelinek plot of input s data.
Example: 5.13
Options: -p parametric bootstrap
Input: s_1 s_2 s_3 s_4 s_5 s_6 $[\sigma]$
Output: plotxy commands for P' versus T diagram

- **s_tern**

Usage: s_tern [-p][Standard I/O]

Description: Makes a Ternary projection of input s data.

Example: See Chapter 5

Options: -p parametric bootstrap

Input: s_1 s_2 s_3 s_4 s_5 s_6 $[\sigma]$

Output: plotxy command file of ternary diagram

Notes: triangles are data, dots are bootstrapped means

- **s_tilt**

Usage: s_tilt [Standard I/O]

Description: Rotates s data into tilt adjusted coordinates

Example: 5.3

Input: s_1 s_2 s_3 s_4 s_5 s_6 strike dip

Output: s_1 s_2 s_3 s_4 s_5 s_6 [in tilt adjusted coordinates]

- **splint**

Usage: splint [-i] [interval] [Standard I/O]

Description: Calculates spline interpolation of input.

Option: -i uses interpolation interval [interval]

Default: [interval] = 1

Input: x y data with monotonic increasing x

Output: interpolated x,y

- **stats**

Usage: stats [standard I/O]

Description: Salculates Gauss statistics for input data

Example: 4.6

Input: single column of numbers

Output: N mean sum σ (%) stderr 95%conf.

Notes: σ is the standard deviation % is σ as percentage of the mean stderr is the standard error and 95% conf.$= 1.96\sigma/\sqrt{N}$

- **sundec**

Usage: sundec [-u] [delta T] [Standard I/O]

Description: Calculates declination from sun compass measurements.

Example: 3.1

Options: -u sets the time difference [delta T] in hours from universal time (e.g. -5 for EST)

Input: latitude longitude year month day hours minutes shadow_angle

Output: D

Notes: positive: North, East; negative: South, West

- **vgp_di**

Usage: vgp_di [Standard I/O]

Description: Transforms VGP to equivalent D, I

Example: 1.6

Input: pole_latitude longitude site_latitude longitude.

Output: $D\ I$

Notes: convention is positive: North, negative: South and positive: East, negative: West

- **vspec**

Usage: vspec [Standard I/O]

Description: Calculates a vector average for multiple measurements of a single specimen at a single treatment step.

Example: 6.2

Input:

 (.mag file format - see mag_dat)

  Sample treatment CSD intensity $D\ I$

Output:

 Sample treatment CSD/R* intensity $\bar{D}\ \bar{I}$

 Unique specimen/treatment data are simply copied.

 R* is the vector resultant (including intensity) over the

 sum of all intensities and ranges from 0 to 1.

Appendix 2

TERMS, ACRONYMS AND CONSTANTS

Symbol	Term and Definitions
	Paleomagnetic acronyms and terms
τ	Relaxation time: Section 2.6.1; equation 2.18
θ_m	Magnetic co-latitude: Section 1.3.3; equation 1.14
$[a_m]$	Magnetic activity: Section 3.10.2
APWP	Apparent polar wander path: Section 6.3
AF	Alternating field demagnetization: Section 3.5
ARM	Anhysteretic remanent magnetization: Section 2.6.4
CRM	Chemical remanent magnetization: Section 2.6.5
D	Declination: Section 1.3.1; equation 1.6
DGRF	Definitive geomagnetic reference field: Section 1.3.3
DRM	Detrital remanent magnetization: Section 2.6.6
GAD	Geocentric axial dipole: Section 1.3
GPTS	Geomagnetic polarity time scale: Section 1.4.2; Section 6.1
IGRF	International geomagnetic reference field: Section 1.3.3
IRM	Isothermal remanent magnetization: Section 2.4.1
M_{eq}	Equilibrium magnetization: Section 2.6.2
MD	Multi-domain: Section 2.5
NRM	Natural remanent magnetization: Section 2.6.7
pARM	Partial anhysteretic remanence: Section 2.6.4
pDRM	Post-depositional detrital remanent magnetization: Section 2.6.6
PSD	Pseudo-single domain: Section 2.5
PSV	Paleosecular variation of the geomagnetic field: Section 1.4.1
pTRM	Partial thermal remanence: Section 2.6.3
sIRM	Saturation IRM: See M_r
SD	Single domain: Section 2.5
SP	Superparamagnetic: Section 2.6.1
SV	Secular variation: Section 1.4.1
TRM	Thermal remanent magnetization: Section 2.6.3
VADM	Virtual axial dipole moment: Section 1.3.5
VDM	Virtual dipole moment: Section 1.3.5; equation 1.19
VDS	Vector difference sum: Section 3.7
VGP	Virtual geomagnetic pole: Section 1.3.4
VRM	Viscous remanent magnetization: Section 2.6.2
	Miscellaneous terms
GHA	Greenwich hour angle: Section 3.1
SQUID	Superconducting quantum interference device: Section 3.4
UT	Universal time (Greenwich mean time): Section 3.1

Physical Parameters and Constants

α	Co-inclination: Section 4.5
χ	Magnetic susceptibility: The slope relating induced magnetization to an applied field: Section 1.1
χ_b	Bulk magnetic susceptibility: Section 5.3; equation 5.23
χ_d	Diamagnetic susceptibility: Section 2.1.1
χ_f	Ferromagnetic susceptibility: Section 2.2; equation 2.11
χ_p	Paramagnetic susceptibility: Section 2.1.2; equation 2.9
ΔM curve	Curve defined by subtracting the ascending from the descending curves in a hysteresis loop: Section 2.4.1
λ, ϕ	Latitude, Longitude
μ_o	Permeability of free space: ($4\pi \times 10^{-7}$ Hm^{-1}): Section 1.1.3
θ	Co-latitude: Section 1.3
Θ	Curie temperature. Section 2.2
a_{ij}	Direction cosines: Section 3.2
a	The radius of the Earth (6.371×10^6 m): Section 1.3.3
AMS	Anisotropy of magnetic susceptibility: Section 5.1
B	Magnetic induction: Section 1.1.2
C	Frequency factor (10^{-10} s^{-1}): Section 2.6.1
D	Declination: Section 1.3.1
g_m^l, h_m^l	Gauss coefficients: Section 1.3.3
H	Magnetic field: Section 1.1.1
H_{cr}	Coercivity of remanence; field required to reduce saturation IRM to zero: Section 2.4.1
H_c	Coercivity; the magnetic field required to change the magnetic moment of a particle from one easy axis to another: Section 2.4
H_s	Saturating field; field required to impart M_s: Section 2.4.1
I	Inclination: Section 1.3.1; equation 1.6
k	Boltzmann's constant (1.381×10^{-23} JK^{-1}): Section 2.1.1
K_i	AMS measurement: Section 5.1
K_u	Constant of uniaxial anisotropy energy: Section 2.4
m	Magnetic moment: Section 1.1.3
m_b	Bohr magneton (9.27×10^{-24} Am^2): Section 2.0
M	Magnetization: Section 1.1.3
M_r	Saturation remanence (also sIRM): Section 2.4.1
M_s	Saturation magnetization; the magnetization measured in the presence of a saturating field: Section 2.4.1
P_l^m	Schmidt polynomials: Section 1.3.3
s	Six elements of χ_{ij}; $s_1 = \chi_{11}, s_2 = \chi_{22}, s_3 = \chi_{33}, s_4 = \chi_{12}, s_5 = \chi_{23}, s_6 = \chi_{13}$: Section 5.1; equation 5.4
T	Absolute temperature (in kelvin)
T_b	Blocking temperature: Section 2.6.3
v	Volume
v_b	Blocking volume: Section 2.6.5

	Statistical parameters
α_{95}	Circle of 95% confidence (Fisher): Section 4.1, equation 4.4
δ	Residual errors for AMS measurements: Section 5.2, equation 5.20
ϵ_{ij}	Semi-angles of Hext uncertainty ellipses: Section 5.3; equation 5.21
κ	Fisher precision parameter: Section 4.1
η, ζ	Semi-angles of bootstrap uncertainty ellipses: Section 4.10; equation 4.16
$\tau, \mathbf{V}$	Eigenvalues and eigenvectors of tensors
k	Estimate of κ: Section 4.1
CSD	Circular standard deviation (Fisher): Section 4.1
dm	Uncertainty in the meridian (longitude) of a paleomagnetic pole: Section 4.1; equation 4.5
dp	Uncertainty in the parallel (latitude) of a paleomagnetic pole: Section 4.1; equation 4.5
F, F_{12}, F_{23}	Significance tests for anisotropy (Hext): Section 5.3; equation 5.22
J	Magnetostratigraphic jackknife parameter: Section 6.1.1
LPA	Linear Perturbation Analysis: Section 5.2
MAD	Maximum angular deviation of principal eigenvector (Kirschvink): Section 3.8.2; equation 3.9
MAD_{plane}	MAD of the pole to a best-fit plane (Kirschvink): Section 3.8.2; Equation 3.10
M_u, M_e	Significance tests for uniform and exponential distributions: Section 4.6
N	Number of samples, specimens or sites
n_f	Number of degrees of freedom
R	Resultant vector length (Fisher): Section 4.2
R_o	Critical value of R for non-random distribution (Watson): Section 4.2; equation 4.6
S_o	Residual sum of squares of errors (Hext): Section 5.2; equation 5.14
T	Orientation tensor: Section 3.8.1; equation 3.6

REFERENCES

Abramowitz, M. and I. Stegun, *Handbook of Mathematical Functions*, National Bureau of Standards, New York, 1970.

Aharoni, A., *Introduction to the Theory of Ferromagnetism*, Oxford University Press, 1996.

Anson, G. L. and K. P. Kodama, Compaction-induced inclination shallowing of the post-depositional remanent magnetization in a synthetic sediment, *Geophys. J. R. astr. Soc.*, *88*, 673–692, 1987.

Backus, G., R. Parker, and C. Constable, *Foundations of Geomagnetism*, Cambridge University Press, 1996.

Balsley, J. and A. Buddington, Magnetic susceptibility anisotropy and fabric of some Adirondack granites and orthogneisses, *Amer. J. Sci.*, *258A*, 6–20, 1960.

Banerjee, S. K., New grain size limits for paleomagnetic stability in hematite, *Nature Phys. Sci.*, *232*, 15–16, 1971.

Banerjee, S. K., J. King, and J. Marvin, A rapid method for magnetic granulometry with applications to environmental studies, *Geophys. Res. Lett.*, *8*, 333–336, 1981.

Bard, E., B. Hamelin, R. G. Fairbanks, and A. Zindler, Calibration of the ^{14}C timescale over the past 30,000 years using mass spectrometric U-Th ages from Barbados corals, *Nature*, *345*, 405–410, 1990.

Barton, C., *International Geomagnetic Reference Field, 1995 revision*, EOS, Trans. AGU., 1996.

Barton, C. E. and M. W. McElhinny, Detrital remanent magnetization in five slowly redeposited long cores of sediment, *Geophys. Res. Lett.*, *6*, 229–232, 1979.

Behrensmeyer, A. K. and L. Tauxe, Isochronous fluvial systems in Miocene deposits of northern Pakistan, *Sedimentology*, *29*, 331–352, 1982.

Bingham, C., *Distributions on the sphere and on the projective plane*, Ph.D. Thesis, Yale University, 1964.

Blakely, R., *Potential Theory in Gravity and Magnetic Applications*, Cambridge University Press, 1995.

Borradaile, G. J., Magnetic susceptibility, petrofabrics and strain, *Tectonophysics*, *156*, 1–20, 1988.

Box, G. and J. Hunter, Multi-factor experimental designs for exploring response surfaces, *Ann. Math. Statist.*, *28*, 195–241, 1957.

Briden, J. C. and M. A. Ward, Analysis of magnetic inclination in bore cores, *Pure Appl. Geophys.*, *63*, 133–152, 1966.

Bullard, E., J. Everett, and A. Smith, A symposium on continential drift - IV. The fit of the continents around the Atlantic, *Phil. Trans. Roy. Soc.*, *258*, 41–51, 1965.

Butler, R. F., *Paleomagnetism*, Blackwell Scientific Publications, 1992.

Cande, S. and D. Kent, Revised calibration of the geomagnetic polarity timescale for the late Cretaceous and Cenozoic, *J. Geophys. Res.*, *100*, 6093–6095, 1995.

Cande, S. C., and D. V. Kent, A new geomagnetic polarity time scale for the late Cretaceous and Cenozoic, *J. Geophys. Res.*, *97*, 13917–13951, 1992.

Cañon-Tapia, E., E. Herrero-Bervera, and G. Walker, Flow directions and paleomagnetic study of rocks from the Azufre volcano, Argentina, *J. Geomag. Geoelectr.*, *46*, 143–159, 1994.

Chapman, S. and J. Bartels, *Geomagnetism*, Oxford University Press, London, 1940.

Chikazumi, S. and S. Charap, *Physics of Magnetism*, Krieger, Melbourne, 1986.

Clement, B. M. and D. V. Kent, A detailed record of the lower Jaramillo polarity transition from a southern hemisphere, deep-sea sediment core, *J. Geophys. Res.*, *89*, 1049–1058, 1984.

Coe, R. S., The determination of paleointensities of the Earth's magnetic field with emphasis on mechanisms which could cause non-ideal behavior in Thellier's method, *J. Geomag. Geoelectr.*, *19*, 157–178, 1967a.

Coe, R. S., Paleointensities of the Earth's magnetic field determined from Tertiary and

References

Quaternary rocks, *J. Geophys. Res.*, *72*, 3247–5281, 1967b.

Coe, R. S., S. Grommé, and E. A. Mankinen, Geomagnetic paleointensities from radiocarbon-dated lava flows on Hawaii and the question of the Pacific nondipole low, *J. Geophys. Res.*, *83*, 1740–1756, 1978.

Collinson, D. W., *Methods in Rock Magnetism and Paleomagnetism*, Chapman and Hall, 1983.

Constable, C. and L. Tauxe, The bootstrap for magnetic susceptibility tensors, *J. Geophys. Res.*, *95*, 8383–8395, 1990.

Constable, C. and L. Tauxe, Towards absolute calibration of sedimentary paleointensity records, *Earth Planet. Sci. Lett.*, *143*, 269–274, 1996.

Constable, C., L. Tauxe, and R. Parker, Analysis of 11 Myr of geomagnetic intensity variation, in press, 1998.

Day, R., M. D. Fuller, and V. A. Schmidt, Hysteresis properties of titanomagnetites: grain size and composition dependence, *Phys. Earth Planet. Int.*, *13*, 260–266, 1977.

Dekkers, M. J., Magnetic properties of natural pyrrhotite. II. High and low temperature behaviors of Jrs and TRM as a function of grain size, *Phys. Earth Planet. Inter.*, *57*, 266–283, 1989a.

Dekkers, M. J., Magnetic properties of natural goethite I. Grain size dependence of some low and high field related rock magnetic parameters measured at room temperature, *Geophys. J.*, *97*, 323–340, 1989b.

Dekkers, M. J., J.-L. Mattei, G. Fillion, and P. Rochette, Grain-size dependence of the magnetic ,behavior of pyrrhotite during its low temperature transition at 34 K., *Geophys. Res. Lett.*, *16*, 855–858, 1989.

Dunlop, D., Developments in rock magnetism, *Rep. Prog. Phys.*, *53*, 707–792, 1990.

Dunlop, D., Magnetism in rocks, *J. Geophys. Res.*, *100*, 2161–2174, 1995.

Dunlop, D. and . Ö. Özdemir, *Rock Magnetism: Fundamentals and Frontiers*, Cambridge University Press, 1997.

Dunlop, D. and S. Xu., Theory of partial thermoremanent magnetization in multidomain grains, 1. Repeated identical barriers to wall motion (single microcoercivity), *J. Geophys. Res.*, *99*, 9005–9023, 1994.

Efron, B., *The Jackknife, the Bootstrap and Other Resampling Plans*, SIAM (Regional Conf. Ser. in App. Math.), 1982.

Ellwood, B. B., F. Hrouda, and J.-J. Wagner, Symposia on magnetic fabrics: Introductory comments, *Phys. Earth Planet. Inter.*, *51*, 249–252, 1988.

Elsasser, W. M., E. P. Ney, and J. R. Wenkker, Cosmic ray intensity and geomagnetism, *Nature*, *178*, 1226, 1956.

Fisher, N. I. and P. Hall, New statistical methods for directional data 1. bootstrap comparison of mean directions and the fold test in palaeomagnetism, *Geophys. J. Int.*, *101*, 305–313, 1990.

Fisher, N. I., T. Lewis, and B. J. J. Embleton, *Statistical Analysis of Spherical Data*, Cambridge University Press, 1987.

Fisher, R. A., Dispersion on a sphere, *Proc. Roy. Soc. London, Ser. A*, *217*, 295–305, 1953.

Flinn, D., On folding during three-dimensional progressive deformation, *Geol. Soc. London Quart. Jour.*, *118*, 385–433, 1962.

Frank, M., B. Schwarz, S. Baumann, P. Kuhik, M. Suter, and M. A, A 200 kyr record of cosmogenic radionuclide production rate and geomagnetic field intensity from ^{10}Be in globally stacked deep-sea sediments, *Earth Planet. Sci. Lett.*, *144*, 121–129, 1996.

Fuller, M., *Methods of Experimental Physics*, vol. 24, A, chap. Experimental methods in rock magnetism and paleomagnetism, pp. 303–465, Academic Press, 1987.

Gee, J. and D. Kent, Magnetic hysteresis in young mid-ocean ridge basalts: dominant cubic anisotropy?, *Geophys. Res. Lett.*, *22*, 551–554, 1995.

Gilder, S. A., R. S. Coe, H. Wu, G. Kuang, X. Zhao, Q. Wu, and X. Tang, Cretaceous and Tertiary paleomagnetic results from southeast China and their tectonic implications,

287

Earth Planet. Sci. Lett., *117*, 637–652, 1993.

Glen, W., *The Road to Jaramillo*, Stanford University Press, 1982.

Graham, J., The stability and significance of magnetism in sedimentary rocks, *J. Geophys. Res.*, *54*, 131–167, 1949.

Granar, L., Magnetic measurements on Swedish varved sediments, *Arkiv. for Geo.*, *3*, 1–40, 1958.

Grommé, C. S., T. L. Wright, and D. L. Peak, Magnetic properties and oxidation of iron-titanium oxide minerals in Alae and Makaopulhi lava lakes, Hawaii, *J. Geophys. Res.*, *74*, 5277–5293, 1969.

Guyodo, Y., and J.-P. Valet, Relative variations in geomagnetic intensity from sedimentary records: the past 200,000 years, *Earth Planet. Sci. Lett.*, *143*, 23–36, 1996.

Haigh, G., The process of magnetization by chemical change, *Phil. Mag.*, *3*, 267–286, 1958.

Hargraves, R. B., Distribution anisotropy: the cause of AMS in igneous rocks?, *Geophys. Res. Lett.*, *18*, 2193–2196, 1991.

Harland, W., R. Armstrong, A. V. Cox, L. E. Craig, A. G. Smith, D. G. H. W. B. Smith, and A. R. L., *A Geologic Time Scale*, Cambridge University Press, 1989.

Hartl, P., and L. Tauxe, A precursor to the Matuyama/Brunhes transition-field instability as recorded in pelagic sediments, *Earth Planet. Sci. Lett.*, *138*, 121–135, 1996.

Hartl, P., L. Tauxe, and C. Constable, Early Oligocene geomagnetic field behavior from DSDP Site 522, *J. Geophys. Res.*, *98*, 19649–19665, 1993.

Hext, G., The estimation of second-order tensors, with related tests and designs, *Biometrika*, *50*, 353–357, 1963.

Hilgen, F. J., Astronomical calibration of Gauss to Matuyama sapropels in the Mediterranean and implication for the geomagnetic polarity time scale, *Earth Planet. Sci. Lett.*, *104*, 226–244, 1991.

Hoffman, K. A. and R. Day, Separation of multi-component NRM: a general method, *Earth Planet. Sci. Lett.*, *40*, 433–438, 1978.

Housen, B., B. van der Pluijm, and E. Essene, Plastic behavior of magnetite and high strains obtained from magnetic fabrics in the Parry Sound shear zone, Ontario Grenville Province, *J. Struct. Geol.*, *17*, 265–278, 1995.

Ising, G., On the magnetic properties of varved clay, *Arkiv. For. Mate., Astr., Och Fys.*, *29*, 1–37, 1942.

Jackson, M., W. Gruber, J. Marvin, and S. K. Banerjee, Partial anhysteretic remanence and its anisotropy: applications and grainsize-dependence, *Geophys. Res. Lett.*, *15*, 440–443, 1988.

Jackson, M., J. P. Craddock, M. Ballard, R. van der Voo, and C. McCabe, Anhysteretic remanent magnetic anisotropy and calcite strains in Devonian carbonates from the Appalachian Plateau, New York, *Tectonphysics*, *161*, 43–53, 1989.

Jackson, M., H.-U. Worm, and S. K. Banerjee, Fourier analysis of digital hysteresis data: rock magnetic applications, *Phys. Earth Planet. Int.*, *65*, 78–87, 1990.

Jelinek, V., The statistical theory of measuring anisotropy of magnetic susceptibility of rocks and its application, Brno, Geophysika,1-88, 1976.

Jelinek, V., Statistical processing of anisotropy of magnetic susceptibility measured on groups of specimens, *Studia Geophys. et geol.*, *22*, 50–62, 1978.

Jelinek, V., Characterization of the magnetic fabric of rocks, *Tectophys.*, *79*, T63–T67, 1981.

Jiles, D., *Introduction to Magnetism and Magnetic Materials*, Chapman and Hall, 1991.

Joffe, I. and R. Heuberger, Hysteresis properties of cubic single-domain ferromagnetic particles, *Phil. Mag.*, *314*, 1051–1059, 1974.

Johnson, C. and C. Constable, Paleosecular variation recorded by lava flows over the past five million years, *Phil. Trans. Roy. Soc. Lond, Series A - Math. Phys. Eng. Sci.*, *354*, 89–141, 1996.

Juárez, M., L. Tauxe, J. Gee, and T. Pick, The intensity of the Earth's magnetic field over the past 160 million years, Nature, 394, 878–881, 1998.

References

Juárez, T., M. Osete, G. Melendez, C. Langereis, and J. Zijderveld, Jurassic magnetostratigraphy of the Aguilon and Tosos sections (Iberian range, Spain) and evidence of a low-temperature Cretaceous overprint, *Phy. Earth Planet. Inter.*, *85*, 195–211, 1994.

Kamerling, M. J., and B. P. Luyendyk, Tectonic rotations of the Santa Monica mountains region, Western Transverse Ranges, California, suggested by paleomagnetic vectors, *Geol. Soc. Amer. Bull.*, *90, Pt. 1*, 331–337, 1979.

Kent, D. V., Post-depostitional remanent magnetization in deep-sea sediment, *Nature*, *246*, 32–34, 1973.

Kent, J. T., The Fisher-Bingham distribution on the sphere, *J.R.Statist. Soc. B.*, *44*, 71–80, 1982.

King, J. W., S. K. Banerjee, and J. Marvin, A new rock magnetic approach to selecting sediments for geomagnetic paleointensity studies: Application to paleointensity for the last 4000 years, *J. Geophys. Res.*, *88*, 5911–5921, 1983.

King, R. F., The remanent magnetism of artificially deposited sediments, *Mon. Nat. Roy. astr. Soc., Geophys. Suppl.*, *7*, 115–134, 1955.

Kirschvink, J. L., The least-squares line and plane and the analysis of paleomagnetic data, *Geophys. J. Roy. Astron. Soc.*, *62*, 699–718, 1980.

Knight, M. and G. Walker, Magma flow directions in dikes of the Koolau Complex, Oahu, determined from magnetic fabric studies, *J. Geophys. Res.*, *93*, 4301–4319, 1988.

Kok, Y., and L. Tauxe, Saw-toothed pattern of relative paleointensity records and cumulative viscous remanence, *Earth Planet. Sci. Lett.*, *137*, 95–99, 1996a.

Kok, Y., and L. Tauxe, Saw-toothed pattern of sedimentary paleointensity records explained by cumulative viscous remanence, *Earth Planet. Sci. Lett.*, *144*, E9–E14, 1996b.

Kono, M., Statistics of paleomagnetic inclination data, *J. Geophys. Res.*, *85*, 3878–3882, 1980.

Lovlie, R., and T. Torsvik, Magnetic remanence and fabric properties of laboratory deposited hematite-bearing red sandstone, *Geophys. Res. Lett.*, *11*, 229–232, 1984.

Lowrie, W., Identification of ferromagnetic minerals in a rock by coercivity and unblocking temperature properties, *Geophys. Res. Lett.*, *17*, 159–162, 1990.

Lowrie, W., W. Alvarez, g. Napoleone, K. Perch-Nielsen, I. Premoli-Silva, and M. Toumarkine, Paleogene magnetic stratigraphy in Umbrian pelagic carbonate rocks: The Contessa sections, Gubbio, *Geol. Soc. Amer. Bull.*, *93*, 414–432, 1982.

Lu, G., and C. McCabe, Magnetic fabric determined from ARM and IRM anisotropies in Paleozoic carbonates, southern Appalachian Basin, *Geophys. Res. Lett.*, *20*, 1099–1102, 1993.

Lu, R., S. K. Banerjee, and J. Marvin, Effects of clay mineralogy and the electrical conductivity of water on the acquisition of depositional remanent magnetization in sediments, *J. Geophys. Res.*, *95*, 4531–4538, 1990.

Lund, S., J. Liddicoat, T. K. Lajoie, and T. Henyey, Paleomagnetic evidence for long-term (10^4 year) memory and periodic behavior in the Earth's core dynamo process, *Geophys. Res. Lett.*, *15*, 1101–1104., 1988.

McCabe, C., R. van der Voo, C. R. Peacor, C. R. Scotese, and R. Freeman, Diagenetic magnetite carries ancient yet secondary remanence in some paleozoic carbonates, *Geology*, *11*, 221–223, 1983.

McElhinny, M., *Paleomagnetism and Plate Tectonics*, Cambridge University Press, 1973.

McElhinny, M. W., Statistical significance of the fold test in paleomagnetism, *Geophys. J. R. astro. Soc.*, *8*, 338–340, 1964.

McFadden, P. L., A new fold test for paleomagnetic studies, *Geophys. J. Int.*, *103*, 163–169, 1990.

McFadden, P. L., and D. L. Jones, The fold test in paleomagnetism, *Geophys. J. R. astr. Soc.*, *67*, 53–58, 1981.

McFadden, P. L., and F. J. Lowes, The discrimination of mean directions drawn from Fisher distributions, *Geophys J. R. astr. Soc.*, *67*, 19–33, 1981.

References

McFadden, P. L., and M. W. McElhinny, The combined analysis of remagnetization circles and direct observations in paleomagnetism, *Earth Planet. Sci. Lett.*, *87*, 161–172, 1988.

McFadden, P.L., and A. Reid, Analysis of paleomagnetic inclination data, *Geophys. J.R. Astr. Soc.*, *69*, 307–319, 1982.

Merrill, R., M. McElhinny, and P. McFadden, *The magnetic field of the Earth: Paleomagnetism, the Core, and the Deep Mantle*, Academic Press, 1996.

Moskowitz, M., B, Methods for estimating Curie temperatures of titanomaghemites from experimenal Js-T data, *Earth Planet. Sci. Lett.*, *52*, 84–88, 1981.

Nagata, T., *Rock Magnetism*, Maruzen, Tokyo, 2nd edn., 1961.

Nagata, T., Y. Arai, and K. Momose, Secular variation of the geomagnetic total force during the last 5000 years, *J. Geophys. Res.*, *68*, 5277–5282, 1963.

Néel, L., Théorie du trainage magnétique des ferromagneétiques en grains fines avec applications aux terres cuites, *Ann. Geophys.*, *5*, 99–136, 1949.

Néel, L., Some theoretical aspects of rock magnetism, *Adv. Phys.*, *4*, 191–243, 1955.

Nye, J., *Physical Properties of Crystals*, Clarendon Press, Oxford, 1957.

Opdyke, N., and J. Channell, *Magnetic Stratigraphy*, Academic Press, 1996.

Opdyke, N., and K. Henry, A test of the dipole hypothesis, *Earth Planet. Sci. Lett.*, *6*, 139–151, 1969.

O'Reilly, W., *Rock and Mineral Magnetism*, Blackie, 1984.

Owens, W., Mathematical model studies on factors affecting the magnetic anisotropy of deformed rocks, *Tectonophys.*, *24*, 115–131., 1974.

Özdemir, Ö., and S. K. Banerjee, High temperature stability of maghemite (γ-Fe_2O_3), *Geophys. Res. Lett.*, *11*, 161–164, 1984.

Özdemir, Ö., D. Dunlop, and B. Moskowitz, The effect of oxidation of the Verwey transition in magnetite, *Geophys. Res. Lett.*, *20*, 1671–1674, 1993.

Pick, T., and L. Tauxe, Characteristics of magnetite in submarine basaltic glass, *Geophys. J. Int.*, *119*, 116–128, 1994.

Pick, T. S., and L. Tauxe, Geomagnetic palaeointensities during the Cretaceous Normal Superchron measured using submarine basaltic glass, *Nature*, *366*, 238–242, 1993.

Press, W. H., B. P. Flannery, S. Teukolsky, and W. T. Vetterling, *Numerical Recipes: the Art of Scientific Computing*, Cambridge University Press, 1986.

Prévot, M., M. A. Mankinen, C. S. Grommé, and R. S. Coe, How the geomagnetic field vector reverses polarity, *Nature*, *316*, 230–234, 1985.

Prévot, M., M. E. M. Derder, M. McWilliams, and J. Thompson, Intensity of the Earth's magnetic field: evidence for a Mesozoic Dipole Low, *Earth Planet. Sci. Lett.*, *97*, 129–139, 1990.

Ramsay, J., *Folding and fracturing of rocks*, McGraw-Hill Book Co., New York, 1967.

Rees, A., The use of anisotropy of magnetic susceptibility in the estimation of sedimentary fabric, *Sedimentology*, *4*, 257–271, 1965.

Rees, A. I., and W. A. Woodall, The magnetic fabric of some laboratory deposited sediments, *Earth Planet. Sci. Lett.*, *25*, 121–130, 1975.

Roberts, A., Magnetic properties of sedimentary greigite (Fe_3S_4), *Earth Planet. Sci. Lett.*, *134*, 227–236, 1995.

Roberts, A., Y. Cui, and K. Verosub, Wasp-waisted hysteresis loops: Mineral magnetic characteristics and discrimination of components in mixed magnetic systems, *J. Geophys. Res.*, *100*, 17909–17924, 1995.

Rochette, P., G. Fillion, J.-L. Mattei, and M. J. Dekkers, Magnetic transition at 30-34 kelvin in pyrrhotite: Insight into a widespread occurrence of this mineral in rocks, *Earth Planet. Sci. Lett.*, *98*, 319–328, 1990.

Scheidegger, A. E., On the statistics of the orientation of bedding planes, grain axes, and similar sedimentological data, *U.S. Geo. Surv. Prof. Pap.*, *525-C*, 164–167, 1965.

Schneider, D. A. and D. V. Kent, Testing models of the Tertiary paleomagnetic field, *Earth Planet. Sci. Lett.*, *101*, 260–271, 1990.

References

Shackleton, N. J., A. Berger, and W. R. Peltier, An alternative astronomical calibration of the lower Pleistocene timescale based on ODP site 677, *Trans. Roy. Soc. Edinburgh: Earth Sciences*, *81*, 251–261, 1990.

Shimizu, Y., Magnetic viscosity of magnetite, *J. Geomag. Geoelec.*, *11*, 125–138, 1960.

Snowball, I. and R. Thompson, A stable chemical remanence in Holocene sediments, *J. Geophys. Res.*, *95*, 4471–4479, 1990.

Spender, M., J. Coey, and A. Morrish, The magnetic properties and Mössbauer spectra of synthetic samples of Fe$_3$S$_4$, *Can. J. Phys.*, *50*, 2313–2326, 1972.

Stacey, F., G. Joplin, and J. Lindsay, Magnetic anisotropy and fabric of some foliated rocks from S.E. Australia, *Geophysica Pur. Appl.*, *47*, 30–40, 1960.

Stacey, F. D. and S. K. Banerjee, *The Physical Principles of Rock Magnetism*, Elsevier, 1974.

Staudigel, H., J. Gee, L. Tauxe, and R. J. Varga, Shallow intrusive directions of sheeted dikes in the Troodos Ophiolite: anisotropy of magnetic susceptibility and structural data, *Geology*, *20*, 841–844, 1992.

Stokking, L. and L. Tauxe, Properties of chemical remanence in synthetic hematite: Testing theoretical predictions, *J. Geophys. Res.*, *95*, 12639–12652, 1990.

Stoner, E. and W. P. Wohlfarth, A mechanism of magnetic hysteresis in heterogeneous alloys, *Phil. Trans. Roy. Soc. Lond.*, *A240*, 599–642, 1948.

Tanaka, H., M. Kono, and H. Uchimura, Some global features of paleointensity in geological time, *Geophys. J. Int.*, *120*, 97–102, 1995.

Tarduno, J., Superparamgnetism and reduction diagenesis in pelagic sediments: Enhancement or depletion?, *J. Geophys. Res. Lett.*, *22*, 1337–1340, 1995.

Tarling, D. and F. Hrouda, *The Magnetic Anisotropy of Rocks*, Chapman and Hall, London, 1993.

Tauxe, L., Sedimentary records of relative paleointensity of the geomagnetic field: Theory and practice, *Rev. Geophys.*, *31*, 319–354, 1993.

Tauxe, L. and Y. Gallet, A jackknife for magnetostratigraphy, *J. Geophys. Res. Lett.*, *18*, 1783–1786, 1991.

Tauxe, L. and P. Hartl, 11 million years of Oliogocene geomagnetic field behaviour, *Geophys. J. Int.*, *128*, 217–229, 1996.

Tauxe, L. and D. V. Kent, Properties of a detrital remanence carried by hematite from study of modern river deposits and laboratory redepostion experiments, *Geophys. J. Roy. astr. Soc.*, *77*, 543–561, 1984.

Tauxe, L. and G. Watson, The fold test: an eigen analysis approach, *Earth Planet. Sci. Lett.*, *122*, 331–341, 1994.

Tauxe, L., P. Tucker, N. Petersen, and J. LaBrecque, The magnetostratigraphy of Leg 73 sediments, *Palaeogeogr. Palaeoclimat. Palaeoecol.*, *42*, 65–90, 1983.

Tauxe, L., C. Constable, L. Stokking, and C. Badgley, The use of anisotropy to determine the origin of characteristic remanence in the Siwalik red beds of northern Pakistan, *J. Geophys. Res.*, *95*, 4391–4404, 1990.

Tauxe, L., N. Kylstra, and C. Constable, Bootstrap statistics for paleomagnetic data, *J. Geophys. Res.*, *96*, 11723–11740, 1991.

Tauxe, L., T. Pick, and Y. Kok, Relative paleointensity in sediments: a pseudo-Thellier approach, *Geophys. Res. Lett.*, *22*, 2885–2888, 1995.

Tauxe, L., T. Herbert, N. Shackleton, and Y. Kok, Astronomical calibration of the Matuyama Brunhes boundary: consequences for magnetic remanence acquisition in marine carbonates and the Asian loess sequences, *Earth Planet. Sci. Lett.*, *140*, 133–146, 1996a.

Tauxe, L., T. Mullender, and T. Pick, Potbellies, wasp-waists, and superparamagnetism in magnetic hysteresis, *J. Geophys. Res.*, *101*, 571–583, 1996b.

Tauxe, L., J. Gee, and H. Staudigel, Flow directions in dikes from AMS data: The bootstrap way, J. Geophys. Res., 103, 17,775–17,790, 1998.

Thellier, E., and O. Thellier, Sur l'intensité du champ magnétique terrestre dans le passé historique et géologique, *Ann. Geophys.*, *15*, 285–378, 1959.

References

van der Voo, R., Paleomagnetism of North America - a brief review, *Paleoreconstruction of the Continents, Geodynamic Series Amer. Geophys. Union, Geol. Soc. Amer.*, *2*, 159–176, 1981, McElhinny, M.W. and Valencio, D.A. (eds.).

van der Voo, R., Phanerozoic paleomagnetic poles from Europe and North-America and comparisons with continental reconstructions, *Rev. of Geophysics*, *28*, 167–206, 1990.

van der Voo, R., *Paleomagnetism of the Atlantic, Tethys and Iapetus Oceans*, Cambridge University Press, 1993.

Wasilewski, P. J., Magnetic hysteresis in natural materials, *Earth Planet. Sci. Lett.*, *20*, 67–72, 1973.

Watson, G. and R. Enkin, The fold test in paleomagnetism as a parameter estimation problem, *J. Geophys. Res. Lett.*, *20*, 2135–2137, 1993.

Watson, G. S., A test for randomness, *Mon. Not. Roy. Astr. Soc.*, *7*, 160–161, 1956.

Wegener, A., *Die Entstehung der Kontinente*, Dover, 1915.

Woodcock, N., Specification of fabric shapes using an eigenvalue method, *Geol. Soc. Amer. Bull.*, *88*, 1231–1236, 1977.

Worm, H.-U., D. Clark, and M. Dekkers, Magnetic susceptibility of pyrrhotite: Grain size, field and frequency dependence, *Geophys. J. Int.*, *114*, 127–137, 1993.

Xu, S. and D. Dunlop, Theory of partial thermoremanent magnetization in multidomain grains 2. Effect of microcoercivity distribution and comparison with experiment, *J. Geophys. Res.*, *99*, 9025–9033, 1995.

Zijderveld, J., A.C. Demagnetization of rocks: Analysis of results, in *Methods in Paleomagnetism*, edited by D. Collinson, K. Creer, and S. Runcorn, pp. 254–286, Elsevier, New York, 1967.

INDEX

ΔM curve, 53, 70

Abramowitz, M., 140
Aharoni, A., 4, 37, 40
Ampère's law, 4, 5
anisotropy
  magnetic susceptibility, 173
    applications, 230
    bootstrap, 188
    data display, 242
    dikes, 237
    discrimination of eigenvectors, 191
    ellipsoid, 176
    ellipsoid shape, 193
    Hext, 181-185
    Jelinek, 185, 186
    lava flows, 239
    linear perturbation analysis, 178
    measurement schemes, 174
    metamorphic rocks, 240
    paleocurrent directions, 230
    tensor, 173
Anson, G.L., 61
antiferromagnetism, 43, 63, 64
apparent polar wander path, 226
Arai diagram, 106

Backus, G., 3, 13
bacterial magnetite, 63
baked contact test, 93
Balsey, J., 198
Banerjee, S.K., 37, 53, 56, 64, 222
Bard, E., 219
Barton, C.E., 13, 61

Behrensmeyer, A.K., 218
Bingham statistics, 124
Bingham, C., 124
Biot-Savart law, 5
bioturbation, 60, 61
Blakely, R., 3
blocking
  temperature, 57, 58
  volume, 59
Bohr magneton, 37, 40, 63
Boltzmann's constant, 39, 40
bootstrap, 140
  definition, 140
  for virtual poles, 145
  parametric, 142
    tensors, 190
    unit vectors, 144
  simple, 142
    tensors, 188
    unit vectors, 142
  test
    common mean, 146
    eigenvalues, 196
    eigenvectors, 191
    fold, 147
    reversals, 146
Borradaile, G., 193
Box, G, 178
Briden, J.D., 130
Brownian motion, 60
Bullard fit, 229
Bullard, T., 229
Butler, R.F., 3, 37, 85

Cañon-Tapia, E., 239
Cande, S.C., 25–28, 221, 222

293

centripetal force, 38
Channell, J.E.T., 3, 24, 26, 215
Chapman, S., 12
Charap, S., 39
Chikazumi, S., 39
chrons, 25
Clement, B.M., 23, 24
Coe, R.S., 107, 108
coercive field, 45
coercivity
    bulk, 51
    of remanence, 51
compaction, 61
component plot, 100
conglomerate test, 93, 128
Constable, C., 30, 110, 187, 188,
    220, 223, 224
continental drift, 228, 229
coordinate systems, 9
    angular, 10
    cartesian, 9, 10
    dike, 191
    geographic, 90, 91
    sample, 90, 91
    tilt adjusted, 91
    transformation, 10, 90, 91
Coulomb's law, 38
Curie temperature, 41, 42, 57, 60
    estimation, 65
      differential method, 66
      extrapolation method, 66
      intersecting tangents, 66
Curie's law, 40
Curie-Weiss law, 42

Day, R., 54, 55, 69, 70, 99
declination, 10, 11
defect ferromagnetism, 64
Dekkers, M., 74
demagnetization, 95
    alternating field, 59, 96
    data display, 99

progressive, 99
thermal, 96
demagnetizing factor, 46
design matrix, 175
DGRF, 13
diamagnetic susceptibility, 38
diamagnetism, 38, 68
dip
    pole, 15
    structural, 91
dipole
    equation, 17
    formula, 17, 30
    model, 9
    moment, 7
direction cosines, 90
distribution anisotropy, 237
distributions
    Bingham, 124
    Fisher, 124, 125
    Kent, 143
domain
    magnetic, 52
    walls, 53, 54
Dunlop, D., 37, 44, 54, 56, 58, 60,
    73, 107

Efron, B., 139, 217
eigenvalues, 102
eigenvectors, 102
    display, 186
Ellwood, B.B., 186
Elsasser, W.M., 219
Enkin, R., 148, 150
equal area projection, 11
equilibrium magnetization, 57, 58
exchange
    energy, 41, 43
    interactions, 39, 41

ferrimagnetism, 43
ferromagnetism, 41–43, 65
    defect, 64

Index

MODERN APPROACHES IN GEOPHYSICS

1. E.I. Galperin: *Vertical Seismic Profiling and Its Exploration Potential.* 1985
   ISBN 90-277-1450-9
2. E.I. Galperin, I.L. Nersesov and R.M. Galperina: *Borehole Seismology.* 1986
   ISBN 90-277-1967-5
3. Jean-Pierre Cordier: *Velocities in Reflection Seismology.* 1985
   ISBN 90-277-2024-X
4. Gregg Parkes and Les Hatton: *The Marine Seismic Source.* 1986
   ISBN 90-277-2228-5
5. Guust Nolet (ed.): *Seismic Tomography.* 1987
   Hb: ISBN 90-277-2521-7; Pb: ISBN 90-277-2583-7
6. N.J. Vlaar, G. Nolet, M.J.R. Wortel and S.A.P.L. Cloetingh (eds.): *Mathematical Geophysics.* 1988   ISBN 90-277-2620-5
7. J. Bonnin, M. Cara, A. Cisternas and R. Fantechi (eds.): *Seismic Hazard in Mediterranean Regions.* 1988   ISBN 90-277-2779-1
8. Paul L. Stoffa (ed.): *Tau-p: A Plane Wave Approach to the Analysis of Seismic Data.* 1989   ISBN 0-7923-0038-6
9. V.I. Keilis-Borok (ed.): *Seismic Surface Waves in a Laterally Inhomogeneous Earth.* 1989   ISBN 0-7923-0044-0
10. V. Babuska and M. Cara: *Seismic Anisotropy in the Earth.* 1991
    ISBN 0-7923-1321-6
11. A.I. Shemenda: *Subduction.* Insights from Physical Modeling. 1994
    ISBN 0-7923-3042-0
12. O. Diachok, A. Caiti, P. Gerstoft and H. Schmidt (eds.): *Full Field Inversion Methods in Ocean and Seismo-Acoustics.* 1995   ISBN 0-7923-3459-0
13. M. Kelbert and I. Sazonov: *Pulses and Other Wave Processes in Fluids.* An Asymptotical Approach to Initial Problems. 1996   ISBN 0-7923-3928-2
14. B.E. Khesin, V.V. Alexeyev and L.V. Eppelbaum: *Interpretation of Geophysical Fields in Complicated Environments.* 1996   ISBN 0-7923-3964-9
15. F. Scherbaum: *Of Poles and Zeros.* Fundamentals of Digital Seismology. 1996
    Hb: ISBN 0-7923-4012-4; Pb: ISBN 0-7923-4013-2
16. J. Koyama: *The Complex Faulting Process of Earthquakes.* 1997
    ISBN 0-7923-4499-5
17. L. Tauxe: *Paleomagnetic Principles and Practice.* 1998   ISBN 0-7923-5258-0

KLUWER ACADEMIC PUBLISHERS – DORDRECHT / BOSTON / LONDON